Texts in Mathematics

Volume 7

Numbers

Texts in Mathematics Series Editor
Dov Gabbay   dov.gabbay@kcl.ac.uk

Numbers

Melvin Fitting
Greer Fitting

ISBN 978-1-84890-335-7

College Publications
Scientific Director: Dov Gabbay
Managing Director: Jane Spurr

http://www.collegepublications.co.uk

Cover designed by Laraine Welch

To our children
Miriam Fitting and Rebecca Fitting,
who saw the writing of this book,
and to our grandchildren,
Constantin Rotundu and Noah Spencer,
who saw its publication.

Preface

Original Preface

(Written by 1990.)

A preface is supposed to explain why you should read the book. Like most prefaces, this one will make more sense after the fact. Nevertheless, here goes.

Most mathematicians believe (rather strongly) that numbers behave in certain well-defined ways. This belief can not be justified by personal experience. No mathematician has 'seen' more than a finite, probably small, collection of numbers. Instead mathematicians justify their beliefs by giving proofs. In practice, this means that certain facts about numbers are accepted as 'obvious', and used in carefully reasoned arguments for the correctness of other facts that are less obvious, or possibly not obvious at all. Since mathematicians generally are concerned to establish the non-obvious, little thought is customarily given as to why the 'obvious' facts are correct.

Now, it is an observation as old as Aristotle that one can not prove something from nothing. One must always begin with some body of 'obvious' facts and proceed from there. In practice, most mathematicians contentedly place hundreds of facts in this 'obvious' category in order to get on with their proper business of discovery and verification of the non-obvious. But at least once in a mathematician's career, it is good to take a sharp look at the status of the 'obvious' facts; and it is probably best to do it early, and get it over with. As we remarked above, it is not possible to do away with all assumptions, even in mathematics. But, one of the great achievements of 19th and early 20th century mathematics was the careful and precise limitation of exactly what a working mathematician must 'accept on faith'. That is, it was discovered what can constitute an irreducible minimum of 'obvious' facts.

It is the purpose of this book to present such an irreducible minimum, and show how most commonly assumed facts about numbers follow directly.

Nonetheless, this book is a bit of a fraud, because like all mathematicians we still assume that some obvious facts are more obvious than others. This is a book about the number systems, so for our purposes we assume as 'sufficiently obvious' a variety of pre-numerical facts. Specifically, we assume, without being too explicit about the matter, several principles about the behavior of sets or collections. Now this material too has been subjected to a similar treatment, also around the turn of the century. Today one can find basic set theory developed from a small number of axioms in many books on elementary set theory. But our book is long enough already, so we elected to omit this material here. For us the issue is: given a variety of 'obvious facts' from set theory, what elementary properties of numbers must one accept in order to logically derive the entire basic framework of mathematics.

It is also the custom for a Preface to present a summary of the material in the book so that a knowledgeable reader can compare the treatment with others available, without the necessity of reading the book itself. Well, once again, here goes.

Each chapter is a mixture of formal developm ent and informal discussion. Loosely, the formal material is supposed to 'justify' many of the things we routinely do in elementary mathematics, while the informal material considers why we may have been doing those things in the first place.

After an introductory chapter, we devote each chapter to a single number system or, in one instance, to a system of notation. Chapter Two presents the counting numbers via the familiar Dedekind-Peano axioms. The so-called induction axiom explicitly mentions sets, but throughout the chapter, only very simple set-theoretic principles are needed in applications of the axiom. Most people take such things for granted, and so do we. There is, however, an optional section in which we present a justification for definitions by recursion, in particular for our definitions of addition, multiplication and exponentiation. The assumed machinery is more elaborate here. In particular, we use functions, and rely on 'obvious' facts about their behavior. If one has some knowledge of axiomatic set theory, it will be clear that the material in this section could easily be turned into a formal treatment. Without a knowledge of axiomatic set theory, the development in the section should still seem extremely plausible. And finally, one could simply skip the section entirely (it is optional, after all), provided one is willing to accept that our definitions of addition, multiplication and exponentiation are meaningful.

Chapter Three treats the system of whole numbers, that is, 0 is added to the counting numbers of Chapter Two. It is here that subtraction and division with remainder are considered.

Chapter Four introduces place-value notation. Partly this is treated for its own sake, but primarily this chapter is included because of the way we have chosen to treat real numbers: as infinite decimals. Since a place-value name is a sequence of digits, this means we have a new mathematical object to deal with: a finite sequence. It is our belief that finite sequences are things as immediately apparent to our intuition as numbers themselves, consequently we do not try to define them away. Rather, we take them as primitive, and characterize them axiomatically. Once again, if one is familiar with axiomatic set theory, it will be clear how to establish that there are objects about which our axioms are true.

Chapter Five considers the system of rational numbers. The treatment is straightforward.

Chapter Six introduces finite decimals. These are not often treated separately, but they are necessary in our approach as a basis for the treatment of real numbers in the following chapter. But also, we feel they have been somewhat slighted by mathematicians generally. Finite decimals are, after all, the numbers the world actually uses in much of its day to day affairs. Also signed finite decimals are perhaps the simplest example of an integral domain that is not a field, and is not simply the integers themselves.

Chapter Seven presents the system of real numbers, considered as infinite decimals. After all, this is how one generally 'explains' real numbers to a beginning mathematics student. But it turns out that this naive notion can actually be made the basis of a rigorous development, and in a relatively simple fashion. We carry things through a proof that square roots exist. In this chapter we need some more mathematical machinery, beyond what earlier chapters relied on. Specifically we need infinite sequences. We describe these informally, and in terms of functions. Once again, a formal treatment will be obvious if one is familiar with axiomatic set theory.

Chapter Eight introduces signed numbers. Since signed numbers could have been introduced at any of several points, this chapter has been written so that it could follow the one on whole numbers, or on rationals, or on reals. A certain amount of algebra, integral domains and fields, is introduced as a convenient means of organizing the presentation. We conclude with a proof that the signed real numbers constitute a complete ordered field.

We do not present complex numbers. One must stop somewhere, and we have chosen to stop while we still have an ordered field to work in. If we went on to complex numbers, we can think of no good reason for not going on to quaternions and octonians, and the present book is long enough already.

Postscript to the Preface

(Written in 2020 by Melvin Fitting.)

This book was largely written in the 1970's and early 1980's by myself, but with much detailed discussion with my then wife Greer (now deceased) for which she deserves an author credit. Around 1990 it was put into LaTeX, and arrived at its present form by February 28, 1990. Except for this Postscript, the book is exactly as it was then. Since 1990 it has simply sat on the shelf, or rather, on the current version of a shelf—it has been on my web page. It was used in a few classes taught by others, but it essentially remained invisible. I recently realized how much time had passed, looked at the book, and saw that it still had interest. It is finally being made publicly available here.

The point of view taken is that of the average working mathematician or mathematics student who wants to verify the basic number facts, in order to feel safe about things ranging from everyday life to the calculus. The issue is not foundations, but rather having enough certainty to get on with things, leaving the rest to someone who might want to explore at a deeper level. It is essentially the state of the art as of 1900. The 20th century showed that this certainty is seriously misleading.

Sets and their general properties are informally discusssed in Chapter 1, Section 1.3: "...in a *full* development, the notion of set itself would need an axiomatic development too. We do not attempt this. Instead we refer you to any book on set theory. We are going to *assume* you know enough about how sets behave to follow our uses of them, which are fairly simple. For example, we will be assuming there really is a set of things, called counting numbers, about which the axioms given in the next chapter are true."

Sets come into things very directly in two important places. The first place is Axiom 5 in Chapter 2, the *Induction Axiom*: "Any collection of counting numbers which includes 1 and is closed under successor is exactly the collection of counting numbers." This assumes we have some grip on what the sets of counting numbers are. But simple so-called cardinality arguments show that most sets of counting numbers have no descriptions (using precise and meaningful notions of 'description' and 'most'). Well, we can get around this by sticking with sets that do have descriptions. In fact we do say, "we will assume that if we have constructed some 'meaningful' sentence, S, using the logical notions ... and the mathematical notions we are developing, then we can talk about the collection of all counting numbers for which S is true. That is, we assume such a set *exists*." This is fine, but Kurt Gödel famously showed, in 1931, that no axiomatization of the counting numbers, under the assumption that induction will only be used with sets specified using sentences that are built using explicit machinery, fully captures the system of counting numbers. What is allowed as machinery can be quite flexible, but the incompleteness of any such axiomatization remains inescapable. Fortunately, we are not trying to capture *all* the truths of arithmetic, but only what might be called the basic everyday ones. For these, the mathematical language we are using is quite sufficient for all the applications of induction we need, and the incompleteness problem does not arise. This does not mean the problem is not there—it is there, and cannot be avoided as one explores more deeply.

The second place sets arise directly is at the end, in Chapter 8, in the very final section, 8.11. We show that the system of signed real numbers constitutes a complete ordered field, Theorem 8.11.1. We also note that any two complete ordered fields are 'intertranslatable', more properly, they are isomorphic. Thus we have captured *the* signed real numbers completely. Again this is misleading, when looked at closely. One of the conditions for being a complete ordered field is that "Any non-empty set of signed reals which has an upper bound has a least upper bound," Theorem 8.10.4. But this talks about *sets* of signed reals. In fact there are more sets of signed reals than there are sets of counting numbers (using a precise and meaningful notion of 'more'). Then the situation is even worse than it was for the counting numbers. The isomorphism result is, properly speaking, relative to our model of set theory. But our understanding of models of set theory is not only deficient in many details, there are results that say it will remain so.

All this sounds quite negative, but I do not want to end this postscript on a discouraging note. Mathematical knowledge is still the most certain kind of knowledge there is. Through careful application of mathematical tools, fundamental limitations on their capabilities have been discovered. Perhaps it is best to simply conclude that our certainty has led to the certain knowledge that modesty is not just a virtue, but an inescapable necessity.

Contents

CHAPTER 1

Background

This memoir can be understood by any one possessing what is usually called good common sense.... But I feel conscious that many a reader will scarcely recognize in the shadowy forms which I bring before him his numbers which all his life long have accompanied him as faithful and familiar friends; he will...become impatient at being compelled to follow out proofs for truths which to his supposed inner consciousness seem at once evident and certain.

—The Nature and Meaning of Numbers,
in *Essays on the Theory of Numbers*,
Richard Dedekind, 1888

Dedekind remarks in his preface that many will not recognize their old friends the natural numbers in the shadowy shapes which he introduces to them. In this, it seems to me, the supposed persons are in the right–in other words, I am one among them.

—*Principles of Mathematics*,
Bertrand Russell, 1903

1.1 What we intend to study

We intend to study the number systems used in mathematics and in everyday life. Specifically, we will study the counting numbers $(1,2,3,4,\dots)$; the whole numbers (counting numbers with 0); the rational numbers (represented by fractions); incidentally the finite decimals; the real numbers (as represented by infinite decimals); and the signed real numbers. Our primary concern will be the logical structure of the number systems: how the commonly accepted facts about numbers can be justified.

1.2 The axiomatic method

"It was easier to know it than to explain why I know it. If you were asked to prove that two and two made four, you might find some difficulty, and yet you are quite sure of the fact."

—said by Sherlock Holmes,
in *A Study in Scarlet*, A. Conan Doyle, 1887

The system of counting numbers is a rich and intricate mathematical structure. How can we go about investigating it? One natural approach is to think of the number system as relating somehow to the way we see the world around us (adding corresponds to putting collections together, for example) and then rely on our experience and intuition of the world. This is historically the way the counting number system came to be invented, and it is the way it is taught to children. But there are two fundamental difficulties with this approach, one concerned with our experience, one with our intuition.

First, there is an infinity of counting numbers, but any person's experience of the world is a limited one: we are finite creatures. How do we know that something we have never experienced before will behave in a relatively familiar fashion?

The second difficulty is with the notion of intuition, what we 'feel to be true.' It is simply that different people have different intuitions, because different people have had different experiences. Let us consider some examples to illustrate this point.

If one has a list of counting numbers to add, will it matter in what order we add them? Will adding from top down, or from bottom up produce the same result? It will, and for most of us it is so much a part of our personal intuition that it never occurs to us to think about it; we take it as a matter of course. Yet it may not be at all obvious to a child learning how to add.

Call a counting number *prime* if it is not 1, and no counting numbers exactly divide it except 1 and itself. Examples are 2,3,5,7,11 and 13. Is the collection of primes finite or infinite? Most people probably have not considered the problem, and so have little real feeling about the answer. But, a very convincing argument can be given that the collection of primes is infinite. People who are familiar with this argument sooner or later incorporate the result into their subconscious, and from then on it influences their intuitive thinking about all mathematical questions.

Let us say two primes are *paired* if they differ by 2. Examples are 3 and 5; 5 and 7; 11 and 13. Is the collection of paired primes finite or infinite? Here even those who have seriously considered the question have no real conviction about the answer. No one's intuition extends so far—yet. (By the way, one test for real intuition or insight is: can you convince others that you are right? Simple guessing is not enough.)

Intuitions differ. Experience is limited. How is certainty possible? Well, it isn't, but at least one can narrow down the region of uncertainty. Let us consider a not entirely frivolous example.

Suppose someone asked you: "Convince me that $2 + 2$ is 4." Possibly you would do so by saying what 2 means, what 4 means, and what $+$ means, no easy thing to do properly. But instead you might argue as follows. "According to the way I use the notions,

1. 4 is the next number after 3, so 4 is $3 + 1$.

2. Similarly 3 is $2 + 1$;

3. 2 is $1 + 1$;

4. and furthermore if I add three numbers, it doesn't matter which two I add first, or, in symbols, $x + (y + z) = (x + y) + z$."

Then you might say, "I don't know what *you* mean by 2 or the other notions, but whatever you mean by them, do you accept 1) – 4) as true? Because if you do, I can argue that $2 + 2$ is 4 as follows. By 3), 2 is $1 + 1$, so $2 + 2$ is $2 + (1 + 1)$. Then by 4), this is the same as $(2 + 1) + 1$. But by 2) this is the same as $3 + 1$, and by 1) this is just 4."

Thus, it doesn't matter what a person means or thinks he means by 1,2,3,4 and $+$; if he accepts 1) – 4), he can be convinced that $2 + 2$ is 4 by simple reasoning.

Now let us carry these ideas to their extreme. Suppose, once and for all, we get together and write down some statements about the counting numbers that we are all willing to accept. Our reasons for accepting them may vary from person to person. But then, after that, we refuse to rely on anything but those basic statements; we deduce things from these basic facts just as above we deduced that $2 + 2$ is 4 from 1) – 4). Then any uncertainty that might exist about whether the 'laws of arithmetic' are correct is narrowed to the question of the correctness of our basic facts, they are all we need take on faith.

These basic facts, whose truth is simply accepted, are called *axioms*. The procedure of beginning with axioms, and deducing other facts from them by logic alone is called the *axiomatic method*. The most famous example of the axiomatic method is the ancient Greek treatment of geometry. Somewhere around 300 BC, Euclid attempted to deduce all the basic facts of geometry from a few simple, obvious axioms. In Chapter Two we will do the same for the system of counting numbers. Our development is based on that of Peano (1858–1932) and Dedekind (1831–1916) each of whom came to essentially this axiomatic development independently.

We have said we will develop the counting numbers using the axiomatic method. What about the other number systems we will study? Will we need fresh axioms each time we move on to a different system? Generally speaking, we will not (there is an exception to this in Chapter 4). We can define each successive number system in terms of those that came before it.

Why then don't we do something similar with the counting numbers? Define them in terms of something simpler? But then how would we understand the behavior of that simpler thing? Define it in terms of something simpler yet? This can not go on forever. We have to *start* somewhere.

This is an instance of a very general problem. All definitions must be in terms of something else. But to define this something else, we must make use of something else again. And so on. Since there are only a finite number of words in our language, eventually we will find ourselves reusing one of the words we are trying to clarify. The conclusion is: we can not define every concept we use; some must be *primitive*, understood as best we can, but not defined.

But if we can't define all the concepts we will be using, how shall we ever begin the development of our subject? This brings us back to the axiomatic method. We simply agree that such and such notions will be primitive; will be left undefined. Our axioms will involve these notions. The axioms are not definitions. They are statements that use our primitive notions, and that we, for whatever reason, accept as true. This is how we begin.

1.3   What we assume

> Since all definitions are effected by means of other terms, every system of definitions which is not circular must start from a certain apparatus of undefined terms... (These) the *primitive* ideas are *explained* by means of descriptions intended to point out to the reader what is meant; but the explanations do not constitute definitions, because they really involve the ideas they explain.
>
> *Principia Mathematica*
> Alfred North Whitehead and Bertrand Russell, 1910

The account we just gave of the axiomatic method, and the way we, and most mathematicians use it, involves a little cheating. Consider, for instance, the following, which might serve as an axiom: *if x is a counting number then $x + 1$ is a counting number.* We may assume the constant 1, the operation $+$, and the notion of counting number are the primitive concepts that this axiom (and presumably it is one of several) is trying to explicate. These are the things the axiom is 'about', and are left undefined.

But, there is still a good deal more to the sentence: the *logical* terms 'if...then' and 'is'. The 'if...then' construction is one of many related ones that come up constantly in mathematics; other common ones are *and, or, not, for every* and *there exists.* How do we define these terms? We have the same problem here that we do with the more 'mathematical' concepts we are trying to explicate. Well, it is possible to treat these *logical* notions themselves as primitive and investigate (though not define) them. Doing so is really a subject in itself, and we recommend you consult a book on formal (or symbolic or mathematical) logic. We are going to *assume* you know how such logical notions behave, and go on from there, though we do give a brief informal discussion in the next section to make sure we are all using the terms the same way. Notice, though, that that discussion really assumes you already understand the terms; for example, we use the word 'or' in explaining how we are going to use the word 'or'. As we said, we will assume you already understand such words, that you already understand the game. The purpose of the next section is just to agree on the ground rules.

Another term that was used in our candidate for an axiom was 'is'. This innocent looking word is really quite a problem. When we said "...x is a counting number..." what we really had in mind was that there was some *collection* or *set* of things, called counting numbers, and x was one of the things in the collection. Thus we were really assuming that you had some understanding of the behavior of sets or collections (they mean the same thing here).

Once again, in a *full* development, the notion of set itself would need an axiomatic development too. We do not attempt this. Instead we refer you to any book on set theory. We are going to *assume* you know enough about how sets behave to follow our uses of them, which are fairly simple. For example, we will be assuming there really is a set of things, called counting numbers, about which the axioms given in the next chapter are true. And (key fact) we will assume that if we have constructed some 'meaningful' sentence, S, using the logical notions mentioned above, and the mathematical notions we are developing, then we can talk about the collection of all counting numbers for which S is true. That is, we assume such a set *exists*. Principles like these, and a few other simple ones, are used over and over, but since this is not a book on set theory, we assume such principles are already part of your intuition, and we go on from there.

There are a few other mathematical concepts that we make use of, but do not define. For instance, in the chapter on rational numbers we use *ordered pairs*. An ordered pair, informally, is a list of two things in a particular order, and the order is important. Likewise in the chapter on real numbers we use *infinite sequences*. An infinite sequence is an unending list of things, one of which comes first, one of which comes second, and so on. We define such things in terms of *functions*, but we do not really *define* functions, just as we do not really define ordered pairs. We assume such things are parts of the mathematical knowledge you bring to this book. Strict characterizations of them can be found in books on set theory.

So, logically speaking, before you read this book, you should read a book on formal logic, and a book on set theory. Pedagogically speaking, however, if you began your mathematical education with the very first principles you would never understand why such things were being investigated. You've gotten this far in this book; you might as well finish it.

1.4   How we use the words

In this section we discuss informally how we use certain logical terms. This is not a formal development. We assume you already understand the words in a general way, and now we are agreeing on the fine points.

We have already discussed *axioms*. They are our basic assumptions. The truth of an axiom is one of the things we agree to accept, for whatever reason. Or, alternately, it can simply be adopted as a working hypothesis, to see what follows, one of the 'rules of the game' so to speak.

A *theorem* is a statement which can be logically deduced from our axioms, and a logical deduction of it from the axioms is its *proof*.

A *lemma* is (generally speaking) a theorem which is of no particular interest in itself, but which may be used to simplify the proof of some other theorem. It may be referred to as a lemma for the proof of theorem such–and–such.

A *corollary* is a theorem which can be proved in one or two lines using some theorem just established (or its proof). It may be referred to as a corollary to theorem such–and–such.

So much for the external terminology. Now for some of the words used within statements and proofs.

First *or*. Whenever a mathematician uses the word 'or', as in "this or that", he means 'and/or', that is, "this or that or both". If he doesn't want both, he will say so.

Next, *implies*. "This implies that" and "if this then that" mean the same thing. The most common way of proving some statement like "this implies that" is to temporarily *assume* 'this' and see if, from it, using the axioms, 'that' can be deduced. Such a deduction constitutes a proof of "this implies that." Another common method of establishing "this implies that" is to assume one can have 'this' but one need not have 'that' also, and show a contradiction follows.

If and only if and *is equivalent to* have the same meaning. "This is equivalent to that" means "this implies that *and* that implies this." To say two statements are equivalent is to say that from each the other can be deduced.

We assume there are things called *sets* or *collections*, and all that matters about a set is what is in it. If C is a collection of things (for instance, the collection of even counting numbers) then the notation $x \in C$ means x is in C (in this case, x is an even counting number). The notation $x \notin C$ means x is not in C (in this case x is not an even counting number). We will also use the convention that $\{x \mid x$ is an even counting number$\}$ names the set of even counting numbers. (This is read, "the set of those x such that x is an even counting number.")

1.5   The use of variables

We will be using 'algebraic' notation, so we should say a little about the subject. For the time being, let us talk only about the system of counting numbers, $1,2,3,4,\ldots$. We all know that if we multiply a number by 2 it is the same as if we added the number to itself. (Actually, this will be proved in the next chapter. But we need an example now, so we anticipate.) Thus we know the following.

$$2 \times 1 = 1 + 1$$
$$2 \times 2 = 2 + 2$$
$$2 \times 3 = 3 + 3$$
$$\vdots$$

Such expressions are called *identities*.

Earlier we expressed a certain fact in English, namely, "multiplying a number by 2 is the same as adding the number to itself." The expression in quotes sums up all the information contained in the infinitely long list of identities above. But one problem with stating such a fact in English is this: the physical appearance of the sentence bears no relation to the physical appearance of the identities it summarizes. It would be useful if it did since then the rules for operating with particular numbers might naturally extend themselves to become rules for operating with general statements about numbers. Is there another way we can summarize the information, but so that it looks more like the identities we listed?

Suppose we said this. "Whatever number we put in the blank of the following expression, the result is a true statement: $2 \times \_ = \_ + \_$." If we replace the blank by 2 we get $2 \times 2 = 2 + 2$. If we replace the blank by 3 we get $2 \times 3 = 3 + 3$, and so on. We now have summed up all the information we wanted in a single sentence, and we have made use of the form the identities had. In fact, we can get any of the identities themselves mechanically, by simply replacing the blank by a number, and we can do this without actually thinking about the meaning of the words 'plus' and 'times'. Incidentally, this illustrates one of the requirements of a good symbolism: it eliminates some of the need for thinking.

There is still a difficulty here which is illustrated by the following. Suppose we say: "If we add a number to 5 and then add a second number to the result, it is the same as if we had first added the two numbers together and added the result to 5." Now, if we try to use the 'blank system', we will need to sort out blanks. We might try something like this. "Whatever number we put in the blank denoted by _ and whatever number we put in the blank denoted by ~ in the following expression, the result is a true statement: $(5 + \_) + \sim = 5 + (\_ + \sim)$."

Well, we rapidly run out of typographical symbols to denote blanks. It has become customary to use letters instead for this purpose. So we might give the information in the first example we discussed above as "Whatever number we replace x by in the following expression, the result is a true statement: $2 \times x = x + x$." Similarly we might give the information in the second example as: "If we replace x by any number and y by any number in the following expression, the result is a true statement: $(5 + x) + y = 5 + (x + y)$." This sort of thing is often said more shortly: "For any number x, $2 \times x = x + x$" or "for any numbers x and y, $(5 + x) + y = 5 + (x + y)$." Expressions like $2 \times x = x + x$ and $(5 + x) + y = 5 + (x + y)$ which are asserted to give true results whenever the letters contained in them are replaced by any numbers we like, are also called *identities*. Sometimes such things are given in quite terse form. For example, "$2 \times x = x + x$ is true for all x" or "$2 \times x = x + x$ is an identity" or "$2 \times x = x + x$ is a law of numbers".

So much for identities. There is a related problem of expressing conveniently some conditions a number must satisfy to aid us in a search for such a number. For example, suppose we want a number satisfying the following condition: "If we multiply the number by itself and then subtract twice the number, we get 3." We are not now asserting an identity. We do not say every number satisfies this condition. We only claim that some number does, and hope to find one. We can restate the problem as follows: "We want a number such that if we replace x by it in the following expression we get a true statement: $x \times x - 2 \times x = 3$." $x \times x - 2 \times x = 3$ is not now called an identity since we do not assert it is true for any number. Rather, we assert it is true for some number, and ask for one or more such numbers to be found. We call $x \times x - 2 \times x = 3$ an *equation* and call finding a suitable number, *solving the equation.*

So the difference between identities and equations is this: we have two expressions, possibly involving letters, with an equality symbol between them. If we assert that any replacement of the letters by numbers produces a true statement, we have an identity. If we only assert some replacement of letters by numbers will produce a true statement we have an equation. Sometimes an unqualified expression, like $x + y = x$, is given and one must infer from context whether it is meant as an identity or an equation. This becomes easier with experience.

The next question is: by what rules can we manipulate algebraic expressions? That is, what are we allowed to do to an equation to try and solve it, and what can we do to an identity to try and produce a new identity. The answer is simple. An algebraic expression is an (incomplete) statement about numbers. If it is an identity, the letters involved could be replaced by any numbers we like. If it is an equation, there are (we hope) some numbers by which we can replace the letters. At any rate, the letters behave exactly as if they *were* numbers. So what rules may we use? Whatever rules the numbers themselves obey. And we are devoting much of the rest of this book to the study of these rules.

Clearly the systems of numbers must obey different rules, or else we couldn't tell them apart. But does this mean there is an algebra corresponding to each system, each with its own set of rules? In fact, it does. If we are dealing with the algebra of the integers, our letters behave as if they were integers and follow those rules. An identity in this algebra 'holds for' all integers. Likewise, if we are dealing with the algebra of the complex numbers, our identities 'hold for' all complex numbers. Indeed, one reason for having such a system as the complex numbers is that there are many equations in the algebra of integers (also rationals and reals) which have no solutions, but which do when we consider them to be equations in the algebra of complex numbers. Different number systems have different algebras, each obeying different rules. High School Algebra is usually the algebra of real numbers, but it sometimes includes the algebra of complex numbers. Curiously, the algebra of integers is much more difficult.

CHAPTER 2

The Counting Numbers

> ... the numerical laws are really not applicable to the external world:
> they are not laws of nature. They are, however, applicable to judgments,
> which are true of things in the external world: they are laws of the laws
> of nature.
>
> —*Grundlagen der Arithmetik* §87
> G. Frege, 1884

2.1 Counting

We wish to study the counting numbers. We intend to use the axiomatic
method. How shall we formulate our axioms? Well, the basic notion of arithmetic is counting. In this section we informally analyze counting. Later we
will try to capture some of our informal analysis in axioms.

People sometimes object to modern mathematics on the ground that it
doesn't deal with real, concrete things, but only with highly abstract and
imaginary objects. Let us point out that all mathematics deals with imaginary objects, even at the level of counting. Suppose we say, "This apple is
red." Red is a property, and we are saying the apple has the property. What is
it that is red? The apple. But suppose instead we say, "There are three apples
on this table." Well, three is a property. The word 'three' here is used as an
adjective just like 'red' above. To what does the property three apply? Not
to any of the apples. Certainly not to the table. It applies to the *collection*
of apples on the table. But the collection of apples is not a thing in the same
sense that an apple is a thing. Such an object must be considered to be an
idealization, a mental construct, not a physical thing. Even at the level of 'two
apples', 'three apples', we are dealing with abstract, ideal entities.

Early in childhood, by listening to our parents and others, we formed an
idea of the uses of the word 'one'. We learned it was correct to say, "There is
one apple on this table," provided there were apples on the table, and whenever
we removed an apple there were no apples left. Similarly we learned how to
use 'two'. We could say, "There are two apples on this table," provided there
were apples on the table and if we removed one of them, one was left. Likewise
we learned the uses of the words 'three', 'four', 'five' and so on.

Now suppose we are given some collection of apples and are asked what number applies to it. The most straightforward thing to do is this. Pick up one of the apples and move it to a new table. This table now has one apple on it. Pick up another apple and move it to the new table. We now have two apples on this second table because if we remove the apple we just put there we have one left, since we had one before we put it there. Now move another apple to the new table. This makes three on it. We continue this way, and when we have all the apples moved, we know how many of them there are.

In fact, we don't usually move the apples physically. We point to an apple, in effect saying, "I could move this first", then to another, "I could move this next", and so on. In short, we point to the apples one after another saying, "one, two, three," and so on. We count the collection.

In a similar way we can count collections of peaches, or chairs, or whatever. We develop an abstract notion of counting, divorced from particular things. Instead of saying, "one apple, two apples, three apples," and so on, we simply say "one, two, three," and so on. Our notion of *how* to count is independent of *what* we count. Now counting things can be seen as a combination of abstract counting, saying the number names, "one, two, three," and so on, and the very concrete act of pointing at one object after another. The number we name when we point to the last object in the collection is the number associated with the collection.

The process of counting, then, consists of two parts. The sequence of counting numbers, 'one', 'two', 'three', etc. And an assignment of these numbers, in order, to the objects we are counting. The assignment, the pointing, is a psychological process, and is not part of the subject matter which concerns us here. We are interested in the abstract sequence of counting numbers itself.

2.2   The axioms

> I regard the whole of arithmetic as a necessary, or at least natural, consequence of the simplest arithmetic act, that of counting, and counting itself as nothing else than the successive creation of the infinite series of positive integers in which each individual is defined by the one immediately preceding; the simplest act is the passing from an already-formed individual to the consecutive new one to be formed. The chain of these numbers forms in itself an exceedingly useful instrument for the human mind; it presents an inexhaustible wealth of remarkable laws...
>
> —*Continuity and Irrational Numbers*,
> Richard Dedekind, 1872

> As for their other assertion, that God's knowledge cannot comprehend things infinite, it only remains for them to affirm, in order that they may sound the depths of their impiety, that God does not know all numbers. For it is very certain that they are infinite since, no matter at what number you suppose an end to be made, this number can be, I

will not say, increased by the addition of one more, but however great
it be, and however vast be the multitude of which it is the rational and
scientific expression, it can still be not only doubled, but even multiplied. Moreover, each number is so defined by its own properties, that
no two numbers are equal. They are therefore both unequal and different
from one another; and while they are simply finite, collectively they are
infinite.

—*The City of God*, §18
St. Augustine, early 5th century

When we count, we say: "one, two, three," and so on. But 'two' is just the
English name for 'one more than one,' and 'three' is just the English name for
'one more than two' and so on. So, avoiding the complications of our language,
when we count, we essentially are saying: "One, and one more, and one more
than that," and so on. Then what is necessary that we may do this? We need
the number 'one', since the counting process must begin somewhere; its *name*
is not important, all that matters is that we start with it. And we need the
notion of 'and one more than that,' in short, let us say 'successor'.

As is customary, we use the symbol '1' for the number we start with when
we count. If we have, somehow, counted up to a number, say x, we need some
way to denote the next number after x, one more than x, the successor of x.
Let us denote it by x^+. We take these as our basic notions: the number 1, and
the passage from x to x^+. So, our first two axioms for the counting numbers
are:

Axiom 1 1 is a counting number.

Axiom 2 If x is a counting number, so is x^+.

Now, in counting we *start* with 1. We would like to have an axiom that
reflects this. Well, if we could start counting somewhere other than 1, and
count our way to 1, 1 would have to be the next number after some other
number. If we say it is not, in effect we have said that if we want 1 to be in
our counting sequence, we must begin with it.

Axiom 3 For any counting number x, $x^+ \neq 1$.

Next we need an axiom which will guarantee that, as we continue to count,
we keep getting different numbers. At no stage do we repeat an earlier number.
Counting does not proceed like this: 1,2,3,4,5,6,7,3,4,5,6,.... Notice that in this
example, 3 comes after both 2 and 7. That is, different numbers have the same
successor. We don't want this to happen.

Axiom 4 If x and y are counting numbers and $x \neq y$, then $x^+ \neq y^+$.

That is, different numbers have different successors. This axiom is logically
equivalent to the following which is often more convenient.

Axiom 4' If x and y are counting numbers and $x^+ = y^+$ then $x = y$.

Before we go on with our axioms, let us show that as far as we have gone they do as we wanted them to do. Let us show that they guarantee we keep getting new numbers as we count.

  a) By Axiom 1, 1 is a counting number.

  b) By Axiom 2, since 1 is a counting number, so is 1^+. And by Axiom 3, $1^+ \neq 1$, so 1 and 1^+ are different.

  c) By Axiom 2, since 1^+ is a counting number, so is $(1^+)^+$. By Axiom 3, $(1^+)^+ \neq 1$. Also we showed in *b)* that $1^+ \neq 1$. Then by Axiom 4, $(1^+)^+ \neq 1^+$. So 1, 1^+ and $(1^+)^+$ are all different.

  d) This continues. We leave the next step to you.

We come now to the final and most complicated of our axioms. Our intention is to capture, in our axioms, the notion of counting. We have listed several basic facts, but we have not yet fully characterized the notion. As we have just seen, so far we can derive that 1, 1^+, $(1^+)^+$, $((1^+)^+)^+$, and so on, are all counting numbers (and are all distinct). We still need to state, somehow, that these are the *only* things that are counting numbers. We need to say that a collection of counting numbers which includes 1, 1^+, $(1^+)^+$, $((1^+)^+)^+$, and so on, actually has every counting number in it; it is exactly the collection of counting numbers.

The problem here is that we can't say directly that a collection of counting numbers includes all of 1, 1^+, $(1^+)^+$, $((1^+)^+)^+$, etc., without either making infinitely many statements, or using the very concept of counting number we are trying to describe. The 'etc.' above conceals infinitely many statements. A paraphrase like "1 followed by any number of $^+$ symbols" makes use of the notion of number. We must find some indirect way of saying it.

Call a collection of counting numbers *closed under successor* if, whatever number it includes, it also includes the successor of that number. That is, whatever number n is, if n is in the collection, so is n^+. Now, suppose we have a collection of counting numbers which

  a) includes 1,

  b) is closed under successor.

Then it must actually include all of 1, 1^+, $(1^+)^+$, etc. For, by *a)*, it includes 1. By *b)*, since it includes 1, it includes 1^+. By *b)* again, since it includes 1^+, it includes $(1^+)^+$. And so on. We now have the paraphrase we wanted: we can say a collection includes all of 1, 1^+, $(1^+)^+$, etc., by simply saying it includes 1 and is closed under successor.

Axiom 5 Any collection of counting numbers which includes 1 and is closed under successor is exactly the collection of counting numbers.

The role of Axiom 5 may, perhaps, be clarified using the following notion. Call a collection of counting numbers *inductive* if

  a) it includes 1

b) it is closed under successor.

Then Axiom 1 and Axiom 2 together say the collection of all counting numbers is inductive. Axiom 5 says it is the *only* inductive collection.

Axiom 5, often called the *induction axiom*, completes our list of axioms. We have spent much time in discussing their background. We have not tried to prove them. We have argued for their plausibility, that is all. Now we adopt them officially. For the rest of the book, they are 'rules of the game.'

We list the axioms again here for convenience.

Axiom 1 1 is a counting number.

Axiom 2 If x is a counting number, so is x^+.

Axiom 3 For any counting number x, $x^+ \neq 1$.

Axiom 4 If x and y are counting numbers and $x \neq y$, then $x^+ \neq y^+$.

Equivalently

Axiom 4' If x and y are counting numbers and $x^+ = y^+$ then $x = y$.

Axiom 5 Any collection of counting numbers which includes 1 and is closed under successor (that is, which is inductive) is exactly the collection of counting numbers.

Exercises

Exercise 2.2.1 Prove that Axiom 4 and Axiom 4' are equivalent.

Exercise 2.2.2 Show $((1^+)^+)^+$ is a counting number, and is different than any of 1, 1^+, and $(1^+)^+$.

2.3 A few lemmas

In this section we use our axioms and prove one or two lemmas. We will need them later on, but what we are primarily interested in is giving some examples of how things are proved.

To make reading easier, from now on we may write x^{++} for $(x^+)^+$, we may write x^{+++} for $((x^+)^+)^+$, and so on.

Lemma 2.3.1 *For every counting number x, $x^+ \neq x$.*

Remark In words, this says no counting number can be its own successor.

The idea of the proof is this. We form a certain collection of counting numbers, call it C, as follows. Put a number x in C if $x^+ \neq x$. Then (this is the heart of the argument) we show C is inductive. Then, by Axiom 5, C must be exactly the collection of counting numbers. Now, x is in C just if $x^+ \neq x$. But every counting number turns out to be in C. So for every counting number x, $x^+ \neq x$. Now the proof itself.

Proof Let C consist of those counting numbers x such that $x^+ \neq x$. By Axiom 3, $1^+ \neq 1$, so $1 \in C$ (recall, this notation means 1 is a member of C).

Next we show C is closed under successor. Well, suppose $n \in C$; we must show that also $n^+ \in C$. Now if $n \in C$, it means $n^+ \neq n$. Then by Axiom 4 $(n^+)^+ \neq n^+$. But this says $x^+ \neq x$ if x is n^+. So $n^+ \in C$. C is closed under successor.

Hence C is inductive. By Axiom 5, C is the collection of counting numbers. So for every counting number x, $x^+ \neq x$. $\square$

This proof is a typical example of the use of Axiom 5. Proofs using this axiom are called *proofs by induction*. Many of the proofs in this chapter will be by induction.

Lemma 2.3.2 *For every counting number x, either $x = 1$ or else, for some y, $x = y^+$.*

Remark In words this says every counting number either is 1 or is the successor of something.

Proof Let C consist of the number 1, together with all the counting numbers x which are the successors of something, that is, for which there is some y with $x = y^+$.

By definition, $1 \in C$.

Next we show C is closed under successor. Suppose $n \in C$, we show that then $n^+ \in C$. Well, n^+ is the successor of something, namely n. So by definition $n^+ \in C$.

C is inductive, so by Axiom 5 every counting number is in C and we are done. $\square$

Exercises

Exercise 2.3.1 Prove: for every counting number x, $x^{++} \neq x$.

Exercise 2.3.2 Call x a *predecessor* of y if $x^+ = y$. Show no counting number can have two predecessors.

Exercise 2.3.3 Show, for every counting number x, either $x = 1$, or $x = 1^+$, or for some y, $x = y^{++}$.

2.4   Addition

In order to make reading easier, we introduce some official abbreviations.

Definition 2.4.1 We use

2	for	1^+
3	for	2^+
4	for	3^+
5	for	4^+
6	for	5^+
7	for	6^+
8	for	7^+
9	for	8^+

Our intention in this section is to define addition. Before we give our official definition we discuss its informal motivation. This is to make it seem reasonable. But, of course, in proofs we may only use the definition itself, not the ideas that led to it.

We might think of $4+3$ as a set of instructions to us, telling us to start with the number 4 and count off the next 3 numbers. More generally, $x + y$ tells us to start at x and count off the next y numbers. In fact, this is pretty much the way children add before they have learned more complicated techniques. But, as it happens, we choose not to make this our official definition of addition. The reasons why are interesting.

There is a difference between counting in the abstract and counting things. By counting various collections of things many times, the human race gradually abstracted the notion of counting. It went from being able to count apples: "One apple, two apples, three apples,..." to counting without objects: "One, two, three,...". As long as we could only count apples, we couldn't count peaches. Once having developed an abstract notion of counting, we could count anything. (Some languages, in fact, have different sets of counting words for different kinds of things. The notion of counting has not been abstracted out yet. There are traces of a similar development in English: we speak of a flock of sheep, but not of a flock of apples, for example.)

Our axioms are intended to capture the *abstract* notion of counting. We have only listed properties of the counting numbers, and nowhere have we discussed the machinery necessary to count *things*. To count things, we need not only the abstract sequence of counting numbers, but also the ability to assign numbers to the things we are counting.

Now, consider again our informal notion of addition. $4 + 3$ tells us to start at 4 and count off 3 more numbers. To make this our official definition, we would have to be able to count some collection of numbers. Specifically, to add 4 and 3, we would need machinery to 'count off' 5,6,7, and assign them the numbers 1,2,3. But so far we have introduced no machinery for this purpose.

It is quite possible to develop the machinery necessary to 'count off' numbers. But it is complicated, and rather beside the point of our development (though see the optional §5). What we do instead is use our informal notions about adding to work out a characterization of + that does not involve counting things. Then we make this our official definition.

Suppose, using our informal notion of adding, we actually try to compute $4 + 3$. We start at 4 and count off the next 3 numbers. Suppose also we had somehow computed $4 + 2$. Then we could make use of that to simplify our computation of $4 + 3$. Now, $4 + 2$ tells us to start at 4 and count off the next 2 numbers. If we do this, then count off yet another number, the result will be counting off 3 numbers after 4. But we are supposing we have already computed $4+2$. Counting off yet one more number is easy, we just get $(4+2)^+$. To compute $4 + 3$, first compute $4 + 2$, then count off one more number.

Similarly, we could easily compute $4+2$, if we already knew what the value of $4 + 1$ was. $4 + 1$ tells us to start at 4 and count off 1 more number. Say we have done this. If we then count off yet another number, we have in effect started at 4 and counted off 2 more numbers. But this can be symbolized as $(4 + 1)^+$.

Finally, it is easy to say what $4 + 1$ is. To compute it we should start at 4 and count off 1 more number. But this is just 4^+. That is, $4 + 1 = 4^+$.

Let us state the above in more general terms. If x and y are counting numbers, $x + y$ tells us to start at x and count off y more numbers. Then $x + 1$ must be x^+. But also, if $x + y$ means we have counted off y more than x, $x + y^+$ means we have counted off 1 more than we did when we computed $x + y$. Thus $x + y^+$ must be $(x + y)^+$. It would seem addition must have the following properties.

a) $x + 1 = x^+$.

b) $x + y^+ = (x + y)^+$.

Now it turns out that these two properties of addition are, in their turn, sufficient to calculate sums. By way of illustration, let us use them to show $2 + 2 = 4$.

By definition of 2, $2 + 2 = 2 + 1^+$. By *b)*, $2 + 1^+ = (2 + 1)^+$, so $2 + 2 = (2 + 1)^+ = (2^+)^+$. Finally, by definition, $(2^+)^+ = 3^+ = 4$.

With this as background, we give our official characterization.

Definition 2.4.2 + is that operation on the counting numbers which meets the conditions

$$\begin{aligned} x + 1 &= x^+ \\ x + y^+ &= (x + y)^+. \end{aligned}$$

Exercises

Exercise 2.4.1 Show $6 = 2^{++++}$.

Exercise 2.4.2 Use Definition 2.4.2 and show:

1. $5 + 3 = 8$,

2. $3 + 5 = 8$.

2.5   Implicit and explicit definitions

The work in this section is designed to fill a serious gap in our definition of addition. It is not essential for your understanding of what follows, and may be omitted.

A *definition* of a term tells us how to get along without it. For example, by 'aardvark' we mean 'blah, blah, blah.' Then whenever we use the word 'aardvark', we could substitute 'blah, blah, blah' instead. Doing so would not change the correctness of what we say, but it would avoid the use of the term 'aardvark' entirely. The role the term 'aardvark' plays is merely one of convenience: it replaces a longer phrase. For a mathematical example consider the sequence of definitions at the beginning of the last section: 2 is 1^+, 3 is 2^+, etc. These introduce the symbol '3' and tell what it means, that is, how to avoid using it, how to translate it away.

Sometimes definitions are called *explicit definitions* to emphasize this. An explicit definition tells precisely how to replace all occurrences of the term being defined by phrases involving other terms altogether.

But the definition of $+$ in the last section is not of this sort. It does not tell us what the operation $+$ is in *other* terms, rather the definition of $+$ involves $+$ again. $[x + y^+ = (x + y)^+]$. It is not an explicit definition. Some would perhaps argue that it is no definition at all.

An *implicit definition* of a term is an indirect characterization. For example. "The murderer is John Smith," is an *explicit characterization.* "The murderer lives on 4th or 5th Avenue, on the 17th floor, and the last three digits of his or her telephone number are 762," may or may not be an *implicit* characterization. If it turns out that only one person meets all the conditions, it is an implicit characterization, otherwise it isn't. We can't tell if it is an implicit characterization of someone by simply looking at it. More must be done.

Before we consider mathematical examples we need to introduce some more machinery. We need the notion of a *function*. Such an object can be defined and developed rigorously in set theory, but a less formal approach is sufficient for our purposes.

Let C be some collection of counting numbers. A rule which assigns a number to each member of C is a *function with domain C*. For example, suppose C consists of 1,2 and 3, and the rule is:

assign to 1 the number 7
assign to 2 the number 5
assign to 3 the number 2.

This constitutes a function with domain C.

When we talk about functions, we give them names. Generally we call a function f or g or F or G or something similar. Suppose we call the function in the example above f. Then we might say the domain of f consists of 1, 2 and 3, and f assigns to 1 the number 7. and so on. This is customarily abbreviated as follows. One writes $f(x)$ for the number f assigns to x. Then, for our example, $f(1) = 7$, $f(2) = 5$ and $f(3) = 2$.

The domain of a function need not be finite. It can, for example, be the entire collection of counting numbers. In this case the rule will have to be specified by some means other than writing out each instance, as we did in the example above. We might say something like this: let g be the function whose domain is the entire collection of counting numbers, given by $g(x) = x^{++}$. Here the rule is specified by a formula telling us how to compute what g assigns to any given number x.

These examples should suffice for the understanding of what follows. Now let us consider some mathematical examples of implicit and explicit definitions. In these, we assume temporarily that $+$ is available, and that it satisfies the conditions given in Section 2.4.

Suppose I say that I am thinking of a function f whose domain is all counting numbers, meeting the condition:

a) $f(n^+) = [f(n)]^+$.

So $f(4) = [f(3)]^+$, $f(5) = [f(4)]^+$, etc. Does this implicitly characterize a unique function? Well, $f(x) = 3 + x$ meets the condition since, for this choice of f,

$$\begin{aligned} f(n^+) &= 3 + n^+ \\ &= [3 + n]^+ \quad \text{(by the properties of } +\text{)} \\ &= [f(n)]^+ \end{aligned}$$

But by a similar argument, $f(x) = 4 + x$ and $f(x) = 5 + x$ also meet the condition. I have not implicitly characterized a function; many meet the condition.

Suppose I say that I am thinking of a function f meeting the conditions:

a) $f(n^+) = [f(n)]^+$
b) $f(1) = 3$
c) $f(2) = 5$.

Does this implicitly characterize a function? Well

$$\begin{aligned} f(2) &= f(1^+) \quad \text{(definition of 2)} \\ &= [f(1)]^+ \quad \text{(by a)} \\ &= [3]^+ \quad \text{(by b)} \\ &= 4 \quad \text{(definition of 4)} \end{aligned}$$

But by c), $f(2)$ should be 5. I have not implicitly characterized a function; none meet the conditions.

Suppose I say that I am thinking of a function f meeting the conditions:

a) $f(n^+) = [f(n)]^+$
b) $f(1) = 3$.

It turns out that this time I have succeeded in implicitly characterizing a function. It is the function *explicitly* characterized by the formula $f(x) = 2 + x$. To demonstrate that this is so we give the following two arguments.

I. The function $f(x) = 2 + x$ does meet conditions a) and b).

Proof If $f(x) = 2 + x$ then

$$\begin{aligned} f(1) &= 2 + 1 \\ &= 2^+ \qquad \text{(by properties of +)} \\ &= 3 \end{aligned}$$

so condition *b)* is satisfied.

Further, if $f(x) = 2 + x$, then

$$\begin{aligned} f(n^+) &= 2 + n^+ \\ &= [2 + n]^+ \qquad \text{(by properties of +)} \\ &= [f(n)]^+ \end{aligned}$$

so condition *a)* is satisfied. □

II. There can't be two different functions meeting conditions *a)* and *b)*.

Proof Suppose f and f' both meet conditions *a)* and *b)*. We show f and f' are really the same function by showing they assign the same value to each counting number.

Let C be the collection of counting numbers x for which $f(x) = f'(x)$.

Since f and f' both meet condition *b)*, $f(1) = 3$ and $f'(1) = 3$, so $f(1) = f'(1)$. Thus $1 \in C$.

Suppose $n \in C$. This means $f(n) = f'(n)$. Taking successors on both sides, $[f(n)]^+ = [f'(n)]^+$. Since both f and f' meet condition *a)*, this says $f(n^+) = f'(n^+)$, so $n^+ \in C$. Thus C is closed under successor.

By Axiom 5, every counting number is in C, so f and f' agree on every counting number. □

We thus have both implicit and explicit definitions of the same function.

Implicit f is that function meeting the conditions

a) $f(n^+) = [f(n)]^+$

b) $f(1) = 3$

Explicit $f(x) = 2 + x$.

This example is typical. The only way we can be sure a candidate for an implicit definition really defines something is by giving some explicit definition and proving it equivalent. An implicit definition may often be more convenient to work with. For example, the implicit version above does not require us to know what addition means. But unless we've shown it to be equivalent to some explicit definition we don't really know it is safe to use. We don't know it defines anything.

From now on we no longer assume we know about +. That is, we drop the temporary assumption we made so that we could present some mathematical examples.

Now, the definition of $+$ given in Section 2.4 is an implicit one. What we do in the rest of this section is produce an equivalent *explicit* definition of $+$. Having done this once and for all, we need never use the explicit version again; we know the implicit characterization is safe.

Though we are, at the moment, concerned with addition, the same work will allow us to prove a general result that will also apply to multiplication and exponentiation when we come to them. So we prove a general theorem about when implicit definitions can be turned into explicit ones. Then we apply it to addition. Please understand, our work here is of a somewhat different character than elsewhere in the chapter. We here make use of intuitive ideas about functions, for example. We need much more than just the counting number axioms. Our aim now is to show the addition conditions of Section 2.4 are meaningful. But it is the conditions themselves that are used elsewhere in the chapter.

Essentially we generalize the example discussed above. Now, the case in which we successfully characterized a function was the one in which we said what it was to be at 1, $f(1) = 3$, and how to calculate what it does at n^+ if we knew what it did at n, $f(n^+) = [f(n)]^+$. For our generalization we will suppose we are told what f is to be at 1, $f(1) = c$ and we are given some *rule*, call it G, telling us how to calculate what f does at n^+ if we knew what f did at $n, f(n^+) = G(f(n))$. For example, above the rule G was: take successor. That is, $G(x) = x^+$. Part of the point of what follows is that *any* rule will do as well. Such conditions always do characterize exactly one function.

Theorem 2.5.1 (on Definition) *Let c be some counting number. Let G be some function whose domain is the entire set of counting numbers. There is exactly one function F whose domain is all counting numbers meeting the conditions $f(1) = c$ and $f(n^+) = G(f(n))$.*

Essentially what we will do is produce an *explicit* definition of a function f and show it is the only function that meets our conditions. But first, some background work.

Definition 2.5.2 x is a *predecessor* of y if $x^+ = y$.

Thus 3 is a predecessor of 4, for example. Axiom 3 says 1 has no predecessor. By Lemma 2.3.2 and Exercise 2.3.2, every counting number other than 1 has exactly one predecessor.

Definition 2.5.3 Let C be a non-empty collection of counting numbers. We say C is *closed under predecessor* if, whatever number it includes, other than 1, it includes the predecessor of that number as well.

Note that then the collection consisting of just the number 1 *is* closed under predecessor.

Definition 2.5.4 We say a function g is *initial* if its domain is closed under predecessor.

Then if g is initial, 1 is in its domain by Exercise 2.5.1. Also if n^+ is in its domain, so is n.

Definition 2.5.5 Let G be a function whose domain is the entire collection of counting numbers. We say a function g is a *G-function* if g is initial and also, for each n, if n^+ is in the domain of g, then $g(n^+) = G(g(n))$. We say g is a *G-function starting at c* if g is a G-function and also $g(1) = c$.

For example, suppose $G(x) = x^+$. Let g have a domain consisting of 1, 2 and 3, and be given by

$$\begin{aligned} g(1) &= 3 \\ g(2) &= 4 \\ g(3) &= 5. \end{aligned}$$

Then g is a G-function starting at 3. (Make sure you see why.)

Lemma 2.5.6 *Suppose g and h are both G-functions starting at c. If x is in the domain of both g and h, then $g(x) = h(x)$.*

Proof Let C be the following collection of counting numbers: put x in C if *a)* x is not in the domain of g or *b)* x is not in the domain of h or *c)* x is in both domains and $g(x) = h(x)$.

If we can show every counting number is in C, it will establish the lemma. Of course, we do this by showing C is inductive.

By Exercise 2.5.1, 1 is in the domains of both g and h. Since both are G-functions *starting at c*, $g(1) = h(1)$ so condition *c)* is met and $1 \in C$. Next, suppose $n \in C$. If n^+ is not in the domain of either g or h, it is automatically in C. Now suppose n^+ is in both domains. Then n is also in both domains (since g and h are initial). But $n \in C$ and for n neither *a)* nor *b)* holds, so it must be that $g(n) = h(n)$. But then $G(g(n)) = G(h(n))$, on taking G of both sides. But g and h are both G-functions, so this gives us $g(n^+) = h(n^+)$. Then $n^+ \in C$.

C is inductive. Now Axiom 5 completes the proof. □

Lemma 2.5.7 *Let c and G be fixed. For each counting number x there is some G-function starting at c that has x in its domain.*

Proof Let C be the following collection of counting numbers. Put x in C if x is in the domain of some G-function starting at c. We show C is inductive, which, by Axiom 4, is enough to establish the lemma.

Let g be the function with only 1 in its domain, and given by $g(1) = c$. Trivially g is a G-function starting at c. Hence $1 \in C$.

Next we show C is closed under successor. Suppose $n \in C$. Then there is some G-function starting at c, with n in its domain; call it g. We have two cases.

case A) n^+ happens to be in the domain of g too. Then automatically $n^+ \in C$.

case B) n^+ is not in the domain of g. Then we define a new function h as follows. The domain of h is to be the domain of g, together with the number n^+. On the domain of g, h is to be the same as g. And on n^+ we set $h(n^+) = G(g(n))$. We claim h is a G-function starting at c too. Since it obviously has n^+ in its domain, then in *case B)* also we will have $n^+ \in C$. Thus C is closed under successor, and the proof will be finished.

Thus we must show h is a G-function starting at c. We recall that we are in *case B)*, and so n is a counting number in the domain of g, but n^+ is not (though n^+ is in the domain of h).

First we show the domain of h is closed under predecessor. Let w be in the domain of h, and suppose w is not 1, so it has a predecessor. We must show its predecessor is also in the domain of h. We have two cases.

First, suppose w is in the domain of g. But g itself is a G-function, hence initial. So the predecessor of w is also in the domain of g, and hence in the domain of h.

Second, suppose w is not in the domain of g. But w is in the domain of h. So w must be n^+. Then the predecessor of w is n which we know is in the domain of g, and hence in the domain of h.

Thus the domain of h is closed under predecessor. This means h is initial.

Next we show h is a G-function. To do this we suppose k^+ is in the domain of h, and we show $h(k^+) = G(h(k))$. And to do this we again consider two cases.

First, suppose $k = n$. Then $k^+ = n^+$ so

$$
\begin{aligned}
h(k^+) &= h(n^+) \\
&= G(g(n)) \quad \text{(def of } h) \\
&= G(h(n)) \quad \text{(def of } h) \\
&= G(h(k)).
\end{aligned}
$$

Second, suppose $k \neq n$. Then $k^+ \neq n^+$ by Axiom 4. Since the domains of g and h differ only on n^+ then since k^+ is in the domain of h and $k^+ \neq n^+$, k^+ must be in the domain of g. Since k^+ is in the domain of g, so is k, but n^+ isn't. So $k \neq n^+$.

Now k^+ is in the domain of g, and g is a G-function, so $g(k^+) = G(g(k))$. Also $k^+ \neq n^+$ and g and h differ *only* on n^+, so $g(k^+) = h(k^+)$. Similarly since $k \neq n^+$, $g(k) = h(k)$. Combining these three equations gives $h(k^+) = G(h(k))$. In either case, $h(k^+) = G(h(k))$. So h is a G-function.

Finally we show h is a G-function *starting at* c. But this is easy. 1 is in the domains of both g and h, since they are both initial. And g and h agree on the domain of g, hence $g(1) = h(1) = c$ since g is a G-function starting at c.

We have shown h is a G-function starting at c, as promised. This completes *case B)* and concludes the proof. □

The two lemmas above show that, for a given c and G, each counting number x is in the domain of some G-function starting at c, and if x is in the domains of more than one, they agree on x. We are now ready for our main argument. What follows is the proof of the Theorem on Definition 2.5.1.

Proof Let c be a given counting number, and let G be some given function whose domain is the entire set of counting numbers. We produce a unique function f, defined on all counting numbers, and meeting the conditions $f(1) = c$ and $f(n^+) = G(f(n))$.

Well define the function f as follows. For any counting number x, $f(x)$ is to be the same as $g(x)$ where g is any G-function starting at c that has x in its domain. The Lemmas above tell us that this does indeed *explicitly* define a function whose domain is the collection of all counting numbers. Now we show it satisfies our conditions, and is the only function that does so.

I. The function f, just defined, meets the given conditions.

Proof of I.

  a) By the definition of f, $f(1)$ is $g(1)$ where g is any G-function starting at c having 1 in its domain. But by definition of G-function starting at c we must have $g(1) = c$. Hence $f(1) = c$.

  b) By the definition of f, $f(n^+)$ is $g(n^+)$ where g is any G-function starting at c having n^+ in its domain. Choose such a g. Now if n^+ is in the domain of g, so is n, since g is initial, and hence by definition of f again, $f(n)$ is $g(n)$. Also by definition of G-function, $g(n^+) = G(g(n))$. It follows that $f(n^+) = G(f(n))$.

  Thus f meets the given conditions.

II. There can't be two functions meeting the given conditions.

Proof of II. Suppose f and f' both meet the conditions. That is, suppose

$$
\begin{aligned}
f(1) &= c \\
f(n^+) &= G(f(n)) \\
f'(1) &= c \\
f'(n^+) &= G(f'(n)).
\end{aligned}
$$

Form a collection C as follows: put $x \in C$ if $f(x) = f'(x)$. We show every counting number is in C and hence f and f' are identical.

  Well, $f(1) = c$ and $f'(1) = c$ so $f(1) = f'(1)$. Hence $1 \in C$.

  Suppose $n \in C$. Then $f(n) = f'(n)$. Hence $G(f(n)) = G(f'(n))$ and thus $f(n^+) = f'(n^+)$. That is, $n^+ \in C$.

  C is inductive, and we are done. $\square$

Now, finally, let us return to the original issue. The definition of addition given in Section 2.4 was not an *explicit* one. Let us show it is equivalent to an explicit one, and then we are on safe grounds using it.

  In Theorem 2.5.1, take G to be the successor function, $G(x) = x^+$. Then the theorem tells us, for each choice of c there is a unique function meeting the conditions

$$\begin{aligned} f(1) &= c^+ &\text{(note well)}\\ f(n^+) &= [f(n)]^+. \end{aligned}$$

Let us denote this function by f_c. Thus, for each choice of c, f_c is that unique function such that

$$\begin{aligned} f_c(1) &= c^+\\ f_c(n^+) &= [f_c(n)]^+. \end{aligned}$$

Explicit Definition of Addition $x + y$ is the number $f_x(y)$.

Exercises

Exercise 2.5.1 Show every collection that is closed under predecessor contains 1. Hint: suppose C is closed under predecessor but doesn't contain 1. Show, by induction, that for each counting number x, $x \notin C$, so C is empty.

Exercise 2.5.2 Show that the operation $+$, as explicitly defined, meets the conditions $x + 1 = x^+$ and $x + y^+ = (x + y)^+$.

Exercise 2.5.3 Show that if the two operations $+$ and $\oplus$ both meet the addition conditions, then they are the same operation. That is, assume

$$\begin{aligned} x + 1 &= x^+\\ x \oplus 1 &= x^+\\ x + y^+ &= (x + y)^+\\ x \oplus y^+ &= (x \oplus y)^+. \end{aligned}$$

Show: for all x and y, $x + y = x \oplus y$. Hint: hold x fixed, do an induction on y.

2.6   Basic properties of addition

> ...the characteristic of erroneous theories is the impossibility of ever foreseeing new facts; whenever such a fact is discovered, those theories have to be grafted with further hypotheses in order to account for them. True theories, on the contrary... are characterized by being able to predict new facts, a natural consequence of those already known. In a word, the characteristic of a true theory is its fruitfulness.
>
> Louis Pasteur, quoted in
> *The Life Of Pasteur*,
> by Rene Vallery-Radot, 1900,
> translated by Mrs. R. L. Devonshire
> page 243.

We recall our definition of addition. It is that operation meeting the conditions

a) $x + 1 = x^+$

b) $x + y^+ = (x + y)^+$

Freed from algebraic notation, condition *a)* says adding 1 is the same as taking successor. Thus, for example, $(4+3)+1 = (4+3)^+$. Similarly, condition *b)* says that in adding two terms, a successor in the second can be moved outside, and vice versa. For example, $3 + (4 + 5)^+ = [3 + (4 + 5)]^+$. Here the two terms are 3 and $4 + 5$. Now we use our definition and establish some basic results about adding.

Theorem 2.6.1 (Associative law for addition)
For any counting numbers, x, y and z, $x + (y + z) = (x + y) + z$.

Remark This says that when we add three numbers, it doesn't matter which two we choose to add first. We may associate them as we please.

Proof Let b and c be any fixed, but arbitrary, counting numbers. Having chosen b and c, we create a collection C of counting numbers as follows: put z in C if $b + (c + z) = (b + c) + z$. We begin by showing C is inductive. In doing so we indicate which part of the definition of addition we are using at each step.

$$
\begin{aligned}
b + (c + 1) &= b + c^+ & \text{(part a)} \\
&= (b + c)^+ & \text{(part b)} \\
&= (b + c) + 1 & \text{(part a)}
\end{aligned}
$$

This says $1 \in C$.

Next, suppose $n \in C$. We show $n^+ \in C$. Well,

$$
\begin{aligned}
b + (c + n^+) &= b + (c + n)^+ & \text{(part b)} \\
&= [b + (c + n)]^+. & \text{(part b)}
\end{aligned}
$$

But $n \in C$. This means

$$
b + (c + n) = (b + c) + n
$$

and this gives us

$$
[b + (c + n)]^+ = [(b + c) + n]^+.
$$

So,

$$
\begin{aligned}
b + (c + n^+) &= [b + (c + n)]^+ \\
&= [(b + c) + n]^+ \\
&= (b + c) + n^+ & \text{(part b)}
\end{aligned}
$$

and this says $n^+ \in C$.

Then C is inductive. By Axiom 5 every counting number is in C. Then for every z, $b + (c + z) = (b + c) + z$. But we have shown this independently of our choice of b and c; they were entirely arbitrary. Hence we have really shown that for any x, y and z, $x + (y + z) = (x + y) + z$. □

What we did in this proof is quite common. In the future we will simply say something like: fix x and fix y; do an induction on z. This is short for: choose a value for x and choose a value for y; having done so, we show the result holds for all z by induction; then since the choice of x and of y was arbitrary, we in fact have the result for all x and all y, as well as all z.

Lemma 2.6.2 *For any counting number x, $1 + x = x^+$.*

Theorem 2.6.3 (Commutativity of addition) *For any counting numbers x and y, $x + y = y + x$.*

Proof Fix x; we do an induction on y. Let C consist of those counting numbers y for which $x + y = y + x$.

By definition, $x + 1 = x^+$. By Lemma 2.6.2, $1 + x = x^+$. Hence $x + 1 = 1 + x$ so $1 \in C$.

Suppose $n \in C$. This means $x + n = n + x$. We refer to this fact as $*$ below. Now,

$$
\begin{aligned}
x + n^+ &= (x + n)^+ & \text{(part b)} \\
&= (n + x)^+ & \text{(by } * \text{)} \\
&= (n + x) + 1 & \text{(part a)} \\
&= n + (x + 1) & \text{(by Theorem 2.6.1)} \\
&= n + (1 + x) & \text{(since } x + 1 = 1 + x \text{)} \\
&= (n + 1) + x & \text{(by Theorem 2.6.1)} \\
&= n^+ + x & \text{(part a)}
\end{aligned}
$$

And this says $n^+ \in C$.

Hence C is inductive, and the theorem is established. □

We have shown addition of counting numbers is associative and commutative. In everyday life when we add a column of numbers we may, to make things easier, add them in groups first, then add the results together, or we may rearrange them in some more convenient way. That we are justified in doing so follows from the two theorems above. For example, let us show $(3 + 4) + (5 + 6) = [(6 + 4) + 3] + 5$. The argument is as follows.

$$
\begin{aligned}
(3 + 4) + (5 + 6) &= (4 + 3) + (5 + 6) & \text{(commutativity)} \\
&= (4 + 3) + (6 + 5) & \text{(commutativity)} \\
&= [(4 + 3) + 6] + 5 & \text{(associativity)} \\
&= [6 + (4 + 3)] + 5 & \text{(commutativity)} \\
&= [(6 + 4) + 3] + 5 & \text{(associativity)}
\end{aligned}
$$

Notice that we have never added more than two numbers at a time. The result of arguments like this, practically, is that we can simply say: add up the numbers 6, 5, 4 and 3, without specifying the order, because the order never matters.

Theorem 2.6.4 (Cancellation law for addition)
Let x, y and z be counting numbers. If $x + z = y + z$, then $x = y$.

It is tempting to go from $x + z = y + z$ to $x = y$ by subtracting z from both sides. We can't do this because we haven't defined subtraction yet. In fact in the next chapter when we do define subtraction, it will turn out that we need this theorem to make sure our definition makes sense.

Proof Do an induction on z. Let us say z *can be cancelled* if it is correct to go from $x + z = y + z$ to $x = y$. We show every counting number can be cancelled. Well, let C consist of those counting numbers that can be cancelled. It is enough to show C is inductive.

First we show $1 \in C$. Suppose $x + 1 = y + 1$. Then $x^+ = y^+$, so by Axiom 4', $x = y$. Thus 1 can be cancelled; $1 \in C$.

Next we show C is closed under successor. Suppose $n \in C$, that is, n can be cancelled. Now suppose $x + n^+ = y + n^+$. Then $(x + n)^+ = (y + n)^+$. By Axiom 4', $x + n = y + n$. But n can be cancelled, so $x = y$. Thus we have cancelled n^+ so $n^+ \in C$.

C is inductive, and we are done. □

Corollary 2.6.5 (Cancellation law for addition)
Let x, y and z be counting numbers. If $z + x = z + y$ then $x = y$.

Exercises

Exercise 2.6.1 Prove Lemma 2.6.2.

Exercise 2.6.2 Show $[(2 + 4) + 6] + 8 = 6 + [4 + (8 + 2)]$.

Exercise 2.6.3 Show $[(2 + 4) + 6] + 8 = (6 + 8) + (2 + 4)$.

2.7 Multiplication

Our definition of multiplication will be along the same lines as our definition of addition, and the general comments we made there apply here too.

Just as addition can be thought of as repeated counting, multiplication can be thought of as repeated addition. Our informal guide is: to multiply 4 by 3 we write down 3 4's and add them. But how can we express this without dragging in machinery to count things? We look for an alternative definition which will result in $4 \cdot 3$ being the result of adding 4, three times, but won't make it necessary to count 4's. We follow the general pattern of the definition of addition; what can we say about multiplying by 1; what can we say about multiplying by y^+ if we know about multiplying by y. We may be guided by the informal notion that $x \cdot y$ means add y x's together.

Well, $x \cdot 1$ should be the result of adding up a collection of one x, that is, just x itself. So we want $x \cdot 1 = x$.

Next, suppose we have computed $x \cdot y$; that is, we have added y x's together. What about $x \cdot y^+$? To compute $x \cdot y^+$ we should add up y^+ x's. We can do this by adding y x's, and then adding in one more x. But adding y x's is just $x \cdot y$, and adding one more x gives us $x \cdot y + x$. So we want $x \cdot y^+ = x \cdot y + x$.

Definition 2.7.1 Multiplication is that operation on the counting numbers meeting the following conditions:

  a) $x \cdot 1 = x$
  b) $x \cdot y^+ = x \cdot y + x$.

The rest of this section is only for those who read Section 2.5.

The definition of multiplication above is an *implicit* one and should be shown to be equivalent to some explicit definition for security. But this is quite easy because of the powerful tool we have in the Theorem on Definition 2.5.1.

Let c be some fixed counting number, and let G be the function given by $G(x) = x + c$. Then by Theorem 2.5.1 there is a unique function meeting the conditions:

$$
\begin{aligned}
f(1) &= c \\
f(n^+) &= G(f(n)).
\end{aligned}
$$

Let us denote this function by f_c. Then, for each c, f_c is that unique function such that

$$
\begin{aligned}
f_c(1) &= c \\
f_c(n^+) &= G(f_c(n)).
\end{aligned}
$$

Explicit Definition of Multiplication $x \cdot y$ is the number $f_x(y)$.

Exercises

Exercise 2.7.1 Use the definition of multiplication to compute:

  1. $4 \cdot 2$,

  2. $2 \cdot 4$.

Exercise 2.7.2 Show $x \cdot 3 = (x + x) + x$.

Exercise 2.7.3 Show that the multiplication operation as explicitly defined meets the conditions:

$$
\begin{aligned}
x \cdot 1 &= x \\
x \cdot y^+ &= x \cdot y + x.
\end{aligned}
$$

Exercise 2.7.4 Show that only one operation can meet the conditions for multiplication given above.

2.8   Basic properties of multiplication

Our definition of multiplication says that it meets the conditions

   a) $x \cdot 1 = x$
   b) $x \cdot y^+ = x \cdot y + x$

Part *a)* says multiplying (on the right) by 1 doesn't change things. For example, $(4 \cdot 3) \cdot 1 = 4 \cdot 3$. Similarly part *b)* says that multiplying (on the right) by y^+ is just multiplying by y, and then throwing in one more x. For example, $(4 + 3) \cdot 5^+ = (4 + 3) \cdot 5 + (4 + 3)$. Now we proceed to establish some basic properties of multiplication.

Theorem 2.8.1 (Right distributive law) *For any counting numbers x, y and z, $(x + y) \cdot z = x \cdot z + y \cdot z$.*

Proof Fix x and y, we do an induction on z. Let C consist of those counting numbers z for which $(x + y) \cdot z = x \cdot z + y \cdot z$. We show C is inductive.
   First, $(x + y) \cdot 1 = x + y = x \cdot 1 + y \cdot 1$, by *a)*. Hence $1 \in C$. Next, suppose $n \in C$. This means

$$(x - y) \cdot n = x \cdot n + y \cdot n \qquad (*)$$

We show $n^+ \in C$. Our argument makes much use of the commutativity and associativity of multiplication. Make sure you see the reason for each step. Now,

$$
\begin{aligned}
(x + y) \cdot n^+ &= (x + y) \cdot n + (x + y) &&\text{(part b)}\\
&= (x \cdot n + y \cdot n) + (x + y) &&\text{(by *)}\\
&= x \cdot n + [y \cdot n + (x + y)]\\
&= x \cdot n + [y \cdot n + (y + x)]\\
&= x \cdot n + [(y \cdot n + y) + x]\\
&= x \cdot n + [x + (y \cdot n + y)]\\
&= (x \cdot n + x) + (y \cdot n + y)\\
&= x \cdot n^+ + y \cdot n^+ &&\text{(part b)}
\end{aligned}
$$

Hence $n^+ \in C$.
   Then C is inductive, and the proof is done. $\square$

Lemma 2.8.2 *For any counting number x, $1 \cdot x = x$.*

Theorem 2.8.3 (Commutative law for multiplication)
For any counting numbers x and y, $x \cdot y = y \cdot x$.

Proof Fix x, we do an induction on y. Let C consist of those counting numbers y for which $x \cdot y = y \cdot x$. We show C is inductive.
   By definition, $x \cdot 1 = x$. By Lemma 2.8.2, $1 \cdot x = x$. Hence $x \cdot 1 = 1 \cdot x$ and this means $1 \in C$.

Next, suppose $n \in C$. This means $x \cdot n = n \cdot x$, which we refer to as $*$. We show $n^+ \in C$. Well,

$$
\begin{aligned}
x \cdot n^+ &= x \cdot n + x & \text{(part b)} \\
&= n \cdot x + x & \text{(by $*$)} \\
&= n \cdot x + 1 \cdot x & \text{(by Lemma 2.8.2)} \\
&= (n+1) \cdot x & \text{(Theorem 2.8.1)} \\
&= n^+ \cdot x
\end{aligned}
$$

so $n^+ \in C$.

C is inductive, and we are done. $\square$

Corollary 2.8.4 (Left distributive law) *For any counting numbers x, y, and z, $x \cdot (y + z) = x \cdot y + x \cdot z$.*

Proof Using commutativity and the right distributive law,

$$
\begin{aligned}
x \cdot (y + z) &= (y + z) \cdot x \\
&= y \cdot x + z \cdot x \\
&= x \cdot y + x \cdot z
\end{aligned}
$$

$\square$

Theorem 2.8.5 (Associative law for multiplication)   *For any counting numbers x, y and z, $x \cdot (y \cdot z) = (x \cdot y) \cdot z$.*

Proof Fix x and y. We do an induction on z. Let C consist of those counting numbers z for which $x \cdot (y \cdot z) = (x \cdot y) \cdot z$. First

$$
\begin{aligned}
x \cdot (y \cdot 1) &= x \cdot y & \text{(part a)} \\
&= (x \cdot y) \cdot 1 & \text{(part a)}
\end{aligned}
$$

so $1 \in C$.

Next, suppose $n \in C$. This means

$$
x \cdot (y \cdot n) = (x \cdot y) \cdot n \quad (*)
$$

We show $n^+ \in C$. Well,

$$
\begin{aligned}
x \cdot (y \cdot n^+) &= x \cdot (y \cdot n + y) & \text{(part b)} \\
&= x \cdot (y \cdot n) + x \cdot y & \text{(Corollary 2.8.4)} \\
&= (x \cdot y) \cdot n + x \cdot y & \text{(by $*$)} \\
&= (x \cdot y) \cdot n^+ & \text{(part b)}
\end{aligned}
$$

and this says $n^+ \in C$.

Hence C is inductive, and we are done. $\square$

Earlier we said that because addition was associative and commutative, we could arrange a sum of several numbers to suit our convenience. We have similar properties for multiplication, and similar results follow.

We have a left and a right distributive law. Generally they are both grouped together under the single name *distributive law*. In high-school algebra language,

$$x \cdot y + x \cdot z = x \cdot (y + z)$$

says we can factor out x from $x \cdot y + x \cdot z$. We will see in Chapter Four that the distributive law is one of the main facts behind the method of multiplication taught in elementary school. And finally, once we have shown that $1 + 1 = 2$, the distributive law gives us an immediate generalization to $x + x = 2 \cdot x$ as follows.

$$\begin{aligned} x + x &= 1 \cdot x + 1 \cdot x \quad &\text{(Lemma 2.8.2)} \\ &= (1 + 1) \cdot x \quad &\text{(distributive law)} \\ &= 2 \cdot x. \end{aligned}$$

Exercises

Exercise 2.8.1 Prove Lemma 2.8.2.

Exercise 2.8.2 Give a proof of the left distributive law, $x \cdot (y + z) = x \cdot y + x \cdot z$, directly from the definition of multiplication, without using any of the results proved in this section.

Exercise 2.8.3 Show $(3 \cdot 4) \cdot (5 \cdot 6) = [(6 \cdot 4) \cdot 3] \cdot 5$.

Exercise 2.8.4 Show $[(2 \cdot 4) \cdot 6] \cdot 8 = 6 \cdot [4 \cdot (8 \cdot 2)]$.

Exercise 2.8.5

1. Show $3 + 2 = 5$.

2. Show $3 \cdot x + 2 \cdot x = 5 \cdot x$.

2.9 Exponentiation

To define the operation of raising to a power, we follow the pattern of our definitions of addition and multiplication.

Multiplication, informally, is repeated addition. We may similarly think of exponentiation as repeated multiplication. We may think of 4^3 as telling us to write down 3 4's and multiply them. And once again we want to express this without involving machinery to count things. Well, x^1 should be the result of multiplying together a collection of one x, that is, just x itself. So we want $x^1 = x$.

Next, suppose we have computed x^y, that is, we have multiplied y x's together. What about x^{y^+}? To compute x^{y^+} we must multiply y^+ x's together. We can do this by first multiplying y x's together, getting x^y, then multiplying by one more x, getting $x^y \cdot x$. So we want $x^{y^+} = x^y \cdot x$.

Definition 2.9.1 Exponentiation is that operation on the counting numbers meeting the conditions

  a) $x^1 = x$
  b) $x^{y^+} = x^y \cdot x$.

Exercises

Exercise 2.9.1 Use the definition of exponentiation to compute:

1. 3^2,

2. 2^3.

Exercise 2.9.2 Show:

1. $x^2 = x \cdot x$,

2. $x^3 = (x \cdot x) \cdot x$.

Exercise 2.9.3 The definition of exponentiation just given is an implicit one. Follow Section 2.5 and:

1. Give an explicit definition of exponentiation;

2. Show your explicitly defined operation meets the exponentiation conditions above;

3. Show only one operation can meet the exponentiation conditions.

2.10   Basic properties of exponentiation

This time we simply state the basic results, and leave the proofs to you.

Theorem 2.10.1 *For any counting numbers x, y and z, $(x \cdot y)^z = x^z \cdot y^z$.*

Theorem 2.10.2 *For any counting numbers x, y and z, $x^{(y+z)} = x^y \cdot x^z$.*

Theorem 2.10.3 *For any counting numbers x, y and z, $(x^y)^z = x^{(y \cdot z)}$.*

Both addition and multiplication are commutative, $x + y = y + x$, and $x \cdot y = y \cdot x$. But exponentiation is not. If you did Exercise 2.9.1 you saw that 3^2 and 2^3 are different. Also both addition and multiplication are associative,

$x + (y + z) = (x + y) + z$ and $x \cdot (y \cdot z) = (x \cdot y) \cdot z$. Again, exponentiation is not.

It is a reasonable question, what makes a fact a *basic* fact. Why did we chose to prove the theorems we did? Our basic operations are addition, multiplication, and exponentiation. Certainly anything that expresses a simple relationship between these operations would be fundamental. What are the possibilities?

Addition and multiplication.

Here the connection is the distributive law, $a \cdot (b + c) = a \cdot b + a \cdot c$. Note that since multiplication is commutative, we needn't state both left and right distributive laws.

Multiplication and exponentiation.

Now, exponentiation is not a commutative operation. So in x^y, we might expect a difference if x or if y is represented as a product. And this is the case. If x is a product, we have Theorem 2.10.1, $(a \cdot b)^c = a^c \cdot b^c$; and if y is the product, we have Theorem 2.10.3, $a^{b \cdot c} = (a^b)^c$.

Addition and exponentiation.

Again, in x^y it makes a difference which of x or y is the sum. We expect two fundamental laws again. If y is the sum, we have Theorem 2.10.2, $a^{(b+c)} = a^b \cdot a^c$. If x is the sum, we have stated nothing that would cover the situation. What can be done with $(a + b)^c$? Actually it is the subject of something called the *Binomial Theorem*. It does not have the simple appearance of the other fundamental laws, but it is as basic as they are. It is, however, beyond the scope of this chapter. We content ourselves with a very special case in an exercise.

Exercises

Exercise 2.10.1 Prove Theorem 2.10.1.

Exercise 2.10.2 Prove Theorem 2.10.2.

Exercise 2.10.3 Prove Theorem 2.10.3.

Exercise 2.10.4 Show that for any counting number x, $1^x = 1$.

Exercise 2.10.5 Evaluate both $2^{(1^2)}$ and $(2^1)^2$, and show they are different.

Exercise 2.10.6 Show $(x + y)^2 = [x^2 + 2(x \cdot y)] + y^2$.

2.11   Order

What does it mean to say 5 is bigger than 2? Intuitively it means that if we start at 2 and count, we will eventually reach 5. In fact, we will reach it after counting off 3 numbers. Thus, $2^{+++} = 5$, and this tells us 5 is greater than 2. More generally, suppose x and y are any counting numbers. What does it mean to say x is greater than y? It means that if we start at y and count we will eventually reach x. How many numbers must we count off? This depends on the relative sizes of x and y. Some undetermined number of them is the best we can say; call it n. Thus $y^{++\cdots+} = x$, where we have 'written' n successor symbols, and this tells us x is greater than y. Now we have seen that $y^{+} = y + 1$, $y^{++} = y + 2$, and $y^{+++} = y + 3$. In fact, for each n, $y^{++\cdots+}$, where we have n $^+$'s, is $y + n$. So to say x is greater than y is to say, for some counting number n, $x = y^{++\cdots+} = y + n$. We will not use an expression like $y^{++\cdots+}$ since it requires us to count things, in this case successor symbols, but $y + n$ is quite acceptable.

With this as motivation, we give our official characterization.

Definition 2.11.1 Let x and y be counting numbers. We say x is greater than y if $x = y + n$ for some counting number n.

If x is greater than y, we write $x > y$, and also $y < x$. Both symbols are due to Harriot, a British mathematician and surveyor of Virginia, who died in 1621. We also write $x \geq y$ for $x > y$ or $x = y$. And $x \leq y$ for $x < y$ or $x = y$.

The two chief results about order are given the names *transitivity* and *trichotomy*. To say $>$ is transitive is to say we can go from $x > y$ and $y > z$ to $x > z$. Let us say x and y are *comparable* if one of $x > y$ or $x = y$ or $x < y$ holds. Trichotomy says that any two counting numbers are comparable, and further, can be compared in only one way. We devote the rest of this section to the proofs of these results.

Theorem 2.11.2 (Transitivity of $>$) *For any counting numbers x, y and z, if $x > y$ and $y > z$ then $x > z$.*

Proof Suppose $x > y$ and $y > z$. Then there are counting numbers n and k such that $y + n = x$ and $z + k = y$. Then

$$
\begin{aligned}
z + (k + n) &= (z + k) + n \\
&= y + n \\
&= x
\end{aligned}
$$

Thus adding $k + n$ to z produces x, so $x > z$. □

The rest of the section is devoted to the proof of the trichotomy law. It is most easily done as a series of small results.

Lemma 2.11.3 *For any counting numbers y and n, $y + n \neq y$.*

Proof Recall, every counting number either is 1 or has a predecessor. We thus can divide the proof into two cases.

Case 1) $y = 1$. Suppose we had $y + n = y$. Then $1 + n = 1$; $n + 1 = 1$, so $n^+ = 1$ contradicting Axiom 3. So $y + n \neq y$.

Case 2) y has a predecessor, say $y = x^+$. Now suppose we had $y + n = y$. Then

$$
\begin{aligned}
x^+ + n &= x^+ \\
(x + 1) + n &= x + 1 \\
x - (1 + n) &= x + 1 \\
x - (n + 1) &= x + 1 \\
x + n^+ &= x + 1 \\
n^+ &= 1
\end{aligned}
$$

The last step above is by the cancellation law for addition. Now we are contradicting Axiom 3 again. Thus $y + n \neq y$. □

Lemma 2.11.4 *Let y be a counting number. We never have $y > y$.*

Proof If $y > y$, then for some n, $y + n = y$, contradicting the previous Lemma.
□

Theorem 2.11.5 *Let x and y be counting numbers. Of the three relationships $x > y$, $x = y$, $x < y$, not more than one can hold.*

Proof
1) Suppose we had $x > y$ and $x = y$. Then, substituting y for x we have $y > y$, contradicting the Lemma above.

2) $x = y$ and $x < y$ are similar.

3) Suppose we had $x > y$ and $x < y$. Since $x < y$, $y > x$. But $x > y$ so by transitivity, $y > y$ again contradicting the Lemma above. □

We now have proved two numbers can not be compared in more than one way. We have yet to show any two numbers can be compared.

Lemma 2.11.6 *For any counting numbers x and y, if $x > y$, then $x \geq y^+$.*

Proof Suppose $x > y$. Then for some n, $x = y + n$. We consider two cases.

case 1) $n = 1$. Then

$$
\begin{aligned}
x &= y + n \\
&= y + 1 \\
&= y^+.
\end{aligned}
$$

Since $x = y^+$, then $x \geq y^+$.

case 2) n has a predecessor say $n = k^+$. Then

$$
\begin{aligned}
x &= y + n \\
&= y + k^+ \\
&= y + (k + 1) \\
&= y + (1 + k) \\
&= (y + 1) + k \\
&= y^+ + k
\end{aligned}
$$

So y^+ with k added gives x. Then $x > y^+$, so again, $x \geq y^+$. □

Let us call a number k a *comparable number* if: for every x, $x > k$ or $x = k$ or $x < k$. That is, k is comparable if the order relationship between it and any counting number is determined.

Lemma 2.11.7 *Every counting number is a comparable number.*

Proof Let C be the collection of counting numbers which are comparable. We show C is inductive.

By Exercise 2.11.3, for every x either $x > 1$ or $x = 1$, hence 1 is a comparable number, $1 \in C$.

Next we show C is closed under successor. Suppose $n \in C$; we show $n^+ \in C$. Let x be any counting number. We must show n^+ and x can be compared. Now $n \in C$, so we have that n and x can be compared, that is, we have one of $n > x$ or $n = x$ or $n < x$. We consider each case separately.

Case 1) $n > x$. By Exercise 2.11.2, $n^+ > n$, so by transitivity, $n^+ > x$. In this case n^+ and x can be compared.

Case 2) $n = x$. By Exercise 2.11.2 again, $n^+ > n$, so $n^+ > x$. Again n^+ and x can be compared.

Case 3) $n < x$. That is, $x > n$. Then by Lemma 2.11.6, $x \geq n^+$, that is, $x > n^+$ or $x = n^+$. Either way, n^+ and x can be compared.

In each case, n^+ and x can be compared. Since x was arbitrary, n^+ is a comparable number, $n^+ \in C$.

Thus C is inductive, and we are done. □

Theorem 2.11.8 (Trichotomy Law) *Let x and y be counting numbers. Exactly one of the following holds: $x > y$, $x = y$, $x < y$.*

Proof By the previous Lemma, x is a comparable number, and so can be compared with y, whatever y is. Thus one of the three relationships must hold. That exactly one holds is by Theorem 2.11.5. □

Exercises

Exercise 2.11.1 Prove the following.

1. If $x > y$ and $y \geq z$ then $x > z$.

2. If $x \geq y$ and $y > z$ then $x > z$.

3. If $x \geq y$ and $y \geq z$ then $x \geq z$.

Exercise 2.11.2 Show for every counting number n, $n^+ > n$.

Exercise 2.11.3 Show for every counting number x, $x \geq 1$. Hint: consider two cases, $x = 1$ and x has a predecessor.

Exercise 2.11.4 Show the equation $x + 7 = 3$ has no solution.

2.12 Insertion and cancellation

We have already proved a cancellation law for addition: if $x + z = y + z$ then $x = y$. Trivially, if $x = y$ then $x + z = y + z$. Putting these into one statement gives us

Theorem 2.12.1 *For any counting numbers x, y and z, $x = y$ if and only if $x + z = y + z$.*

What we do in this section is establish similar results for multiplication and exponentiation, and also similar results with $=$ replaced by $>$.

Theorem 2.12.2 *For any counting numbers x, y and z, $x > y$ if and only if $x + z > y + z$.*

Proof

1) Suppose $x > y$. Then for some counting number n, $x = y + n$. Then

$$
\begin{aligned}
x + z &= (y + n) + z \\
&= y + (n + z) \\
&= y + (z + n) \\
&= (y + z) + n
\end{aligned}
$$

and this says $x + z > y + z$.

2) Suppose $x + z > y + z$. We claim we must have $x > y$. For if we did not, by the trichotomy law we must have one of $x = y$ or $y > x$. If we had $x = y$, we would have $x + z = y + z$, but we have $x + z > y + z$ and the trichotomy law says we can't have both. If we had $y > x$, then by what we just showed in part 1, we would have $y + z > x + z$, and the trichotomy law says we can't have this. So we must have $x > y$. □

Theorem 2.12.3 *For any counting numbers x, y and z, $x > y$ if and only if $x \cdot z > y \cdot z$.*

Proof

1) Suppose $x > y$. Then for some counting number n, $x = y + n$. Then $x \cdot z = (y + n) \cdot z$. Using the distributive law, $x \cdot z = y \cdot z + n \cdot z$, and this says $x \cdot z > y \cdot z$.

2) Suppose $x \cdot z > y \cdot z$. If we had $x = y$, we would have $x \cdot z = y \cdot z$, contradicting the trichotomy law. If we had $y > x$, then by part 1 we would have $y \cdot z > x \cdot z$, again contradicting the trichotomy law. So, using the trichotomy law one more time, we must have $x > y$. □

Theorem 2.12.4 *For any counting numbers x, y and z, $x = y$ if and only if $x \cdot z = y \cdot z$.*

Theorem 2.12.5 *Suppose $x > a$ and $y > b$. Then*

   1. $x + y > a + b$;

   2. $x \cdot y > a \cdot b$.

Proof $x > a$, so by Theorem 2.12.2, $x + y > a + y$. Also $y > b$ so similarly $a + y > a + b$. Then by transitivity, $x + y > a + b$. Part 2 is similar. $\square$

Theorem 2.12.6 *For any counting numbers x, y and z, $x > y$ if and only if $x^z > y^z$.*

Theorem 2.12.7 *For any counting numbers x, y and z, with $z \neq 1$, $x > y$ if and only if $z^x > z^y$.*

Theorem 2.12.8 *For any counting numbers x, y and z:*

   1. $x = y$ *if and only if* $x^z = y^z$;

   2. if $z \neq 1$, $x = y$ *if and only if* $z^x = z^y$.

Exercises

Exercise 2.12.1 Give another proof of part 2 of Theorem 2.12.2 without using the trichotomy law.

Exercise 2.12.2 Prove Theorem 2.12.4.

Exercise 2.12.3 Prove Theorem 2.12.6. Hint: First show that $x > y$ implies $x^z > y^z$ by induction on z, using Theorem 2.12.5. Show the converse by using the trichotomy law.

Exercise 2.12.4 Show that if $z \neq 1$, $z^n > 1$.

Exercise 2.12.5 Prove Theorem 2.12.7. Hint. For the implication from left to right, Theorem 2.10.2 is needed.

Exercise 2.12.6 Prove Theorem 2.12.8.

2.13   The Well Ordering Theorem

Whenever one adds fractions, one puts them over a common denominator. Many of us were taught to use the *least* common denominator. How do we know there is one? More precisely, why must the set S of common denominators have a least member?

    Suppose we are discussing some particular infinite decimal, which we know to be non-zero. Then it has non-zero decimal places. Consider the first non-zero decimal place. But again, how do we know there is one? More precisely, let S be the following collection of counting numbers: put n in S if the nth decimal place is not zero. Why must S have a least member?

Both of these are special cases of the following very general result about the counting numbers.

Theorem 2.13.1 (Well-ordering theorem)
Let S be any non-empty collection of counting numbers. S has a least member.

Remark x is *least* in S if $x \in S$ and, for any $y \in S$, $x \leq y$. The theorem really says something about the order relation of the counting numbers. In general, an order relation is called a *well-ordering* provided any non-empty collection has a least member (in that ordering). Hence the name of the theorem.

Proof Let S be a collection of counting numbers, and suppose S has *no* least member. We show S must be empty. (And so, if S isn't empty after all, it must have had a least member.)

We form a collection C of counting numbers as follows. Put n in C if every counting number $\leq n$ is *outside* S. We show C is inductive.

$x \geq 1$ for every counting number x. It follows that if 1 were in S it would be a least member. But S has no least member so $1 \notin S$. Further since $x \geq 1$ for all x, the only counting number ≤ 1 is 1 itself, and as we just saw, that is outside S. Then by definition, $1 \in C$.

Next we show C is closed under successor. Suppose $n \in C$. We show $n^+ \in C$.

Suppose we had $n^+ \in S$. We claim n^+ would be a least member. For, take any $x \in S$. We can't have $x \leq n$, since $n \in C$, which means all counting numbers $\leq n$ are outside S, but $x \in S$. Since we don't have $x \leq n$, by trichotomy, $x > n$, and so $x \geq n^+$. Since x is any member of S, this says n^+ is a least member of S. But S has no least member. Thus $n^+ \notin S$.

Now suppose $x \leq n^+$. Then either $x = n^+$ or $x < n^+$. By Exercise 2.13.2, either $x = n^+$ or $x \leq n$. If $x = n^+$, $x \notin S$ by the paragraph above. If $x \leq n$, $x \notin S$ since $n \in C$. Hence for any $x \leq n^+$, x is outside S, so by definition, $n^+ \in C$.

Thus C is inductive. Then every counting number is in C. This easily implies S is empty, and the proof is finished. $\square$

As an illustration of how the Well Ordering Theorem is used, we show the following.

Theorem 2.13.2 *There are no counting numbers between 1 and 2.*

Proof Suppose there were counting numbers between 1 and 2. That is, suppose there were counting numbers x with $1 < x < 2$. Let S be the collection of all such. By our supposition, S is not empty, so by the Well Ordering Theorem, S has a least member, call it c. Then $1 < c < 2$, and c is the smallest counting number meeting this condition.

Now, $1 < c$, so $c \neq 1$. Then c must have a predecessor, say $c = d^+$, so that $d \geq 1$. We consider both cases separately.

   case 1) $d = 1$. Then $d^+ = 1^+$, which says $c = 2$. This case is impossible, since $c < 2$.

   case 2) $d > 1$. Now $d < d^+ = c < 2$, so $d < 2$. Then d is also between 1 and 2. Further, $d < d^+ = c$, so $d < c$. But c was the smallest counting number between 1 and 2; d is smaller than c, and is also between 1 and 2. This case is impossible too.

   We have reached contradictions in both cases. The conclusion is, we started out wrong. There are no counting numbers between 1 and 2.   □

Corollary 2.13.3 *There are no counting numbers between a and a^+.*

Proof Suppose we had an x with $a < x < a^+$. Since $x > a$, by definition, $x = a + n$ for some n. Thus what we have is $a < a + n < a^+$. We consider two cases.

   case 1) $a = 1$. Then we are contradicting the Theorem.

   case 2) $a \neq 1$. Then a has a predecessor, say $a = b^+$. Then what we have is $b^+ < b^+ + n < (b^+)^+$. From this we easily get $b + 1 < b + (1 + n) < b + 2$ and by Theorem 2.12.2 we conclude $1 < 1 + n < 2$ again contradicting the Theorem above.

   Thus there are no counting numbers between a and a^+.   □

The Well Ordering Theorem talks about *least* members of non-empty sets of counting numbers. We can not expect a similar result about *greatest* members since, for example, the collection of all counting numbers doesn't have a greatest member. But we do have the following.

Definition 2.13.4 We say a collection C of counting numbers is *bounded by y* if every member of C is $< y$. We say C is *bounded* if it is bounded by some y.

Theorem 2.13.5 *Let C be a bounded, non-empty collection of counting numbers. C has a greatest member.*

Our axioms were about counting. They took as basic the notion of successor. For other purposes, axioms for the counting numbers are sometimes given which take the ordering relation, $>$, as basic. A typical axiom in this approach is: $>$ is transitive. Generally, in this approach, the Well–Ordering Theorem itself is taken as an axiom, and plays a role rather similar to that which Axiom 5, the induction axiom, played in our development. In such a development, our Axiom 5 would be a theorem.

Exercises

Exercise 2.13.1 Show a collection of counting numbers can't have two least members.

Exercise 2.13.2 Suppose $x < n^+$. Show $x \leq n$.

Exercise 2.13.3 Prove Theorem 2.13.5. Hint:

1. Form a collection S as follows. Put y in S if every member of C is $< y$.

2. Show S has a least member, say c.

3. Show $c \neq 1$.

4. Then c has a predecessor, say $c = d^+$. Show $d \in C$ (use the fact that $x < d^+$ implies $x \leq d$).

5. Show d is the greatest member of C.

2.14   Conclusion

We have now developed the basic properties of the counting numbers, and of addition, multiplication, and exponentiation for them. We have said nothing about the inverse operations, subtraction, division and the extraction of roots. It is more convenient to leave subtraction and division for the next chapter. And extraction of roots simply has no natural place in any of the number systems we discuss before we get to the real number system starting in Chapter Seven.

CHAPTER 3

The Whole Number System

Master. If number were so vile a thing as you did esteem it, then need it not be used so much in mens communication. Exclude number, and answer to this question How many years old are you?

Scholar. Mum.

Master. How many dayes in a weeke? How many weeks in a year? What lands hath your Father? How many men doth hee keep? How long is it since you came from him to me?

Scholar. Mum.

Master. So that if number want, you answer all by Mummes: How many miles to London?

Scholar. A poak full of plums.

Master. Why, thus you see, what rule number beareth, and that if number bee lacking it maketh men dumb, so that to most questions they must answer Mum.

−The Ground of Arts,
Chapter *The Declaration of the Profit of Arithmeticke*
Robert Recorde, 1543

3.1   The number zero

For the ancient world, numbers began with 1. There was nothing that played the role of our 0. But eventually people took the natural step and evolved a system of number names based on the operation of the abacus, which in turn was based on finger counting. Now, on an abacus, if we have three beads on a wire, we can represent this by writing down the symbol '3'. If there are no beads on a wire, we need some way of representing this fact too, and at some point the symbol '0' was introduced. Having been introduced for such a purpose, its uses gradually broadened. It began to be treated in much the same way the symbols '1', '2', '3', etc., were. It entered into calculations. The number concept itself broadened in men's minds, so that '0', from being a useful symbol, became the name of a number just as the other symbols were. In short, '1', '2', '3', etc., name numbers and they, or their equivalents, always have. That there is something called a *number* for '0' to name is a relatively recent development. A closer look is in order.

We use the symbol '2' as a name for a certain counting number. This counting number is a mental concept, an idea; specifically, the idea of *twoness*. By considering a great many cases in which it was correct to say, "something, and something else, and that's all," we formed the idea of twoness. Similarly '3' denotes a concept of *threeness* and so on. It is a genuine extension of this group of ideas to include a concept of *nothingness*. By considering many situations in which we have nothing to count, we can abstract a general idea of nothingness. This concept is certainly related to our ideas of oneness, twoness, etc., but it has obvious differences too. It represents our inability to count, rather than the results of our counting. Still, it is natural to extend the number concept to include it. We use '0' to name this concept of nothingness, just as we use '2' to name twoness.

For the rest of this chapter we develop the properties of this new number, denoted by '0'.

3.2   The whole number system

> If zero is added to a number, the sum is that number itself; if zero is subtracted, the number remains unchanged; if zero is multiplied, the result is zero; and if a number is multiplied by zero, the product is zero only.
>
> Sridhara (the Learned)
> Hindu mathematician,
> c. 870 c. 930
> (or perhaps earlier)

Definition 3.2.1 The whole numbers are the counting numbers together with 0 (which is not a counting number).

Next, we use our discussion of the previous sections and come up with a natural extension of the basic operations of arithmetic to include 0. First, counting.

Recall, going from x to x^+ corresponded to counting off one more. If we have a collection of objects and we have counted off 3 of them; after counting off one more, we have counted off 3^+ of them. At that stage at which we have not yet begun to count objects, we may say we have counted off 0 objects. Then, having counted off one object, of course we have counted 1 of them. Thus the successor of 0 ought to be 1.

Definition 3.2.2 $0^+ = 1$.

Next, addition. The intuitive idea on which we based our axioms for counting number addition was: $3 + 2$ tells us to start at 3 and count off the next 2 numbers. Then, applying the same idea, $3 + 0$ tells us to start at 3 and count off the next 0 numbers; that is, don't count. Of course we stay at 3. Then, it is natural to take $3 + 0$ to be 3.

On the other hand, $0 + 3$ tells us to start at 0 and count off the next 3 numbers. That is, $0 + 3$ should be 0^{+++}. But, using our definitions, $0^{+++} = 1^{++} = 2^{+} = 3$. Thus $0 + 3$ also should be 3.

Definition 3.2.3 Let x be any whole number. We take $x + 0 = 0 + x = x$.

We go on to multiplication. The idea we followed in the last chapter was that multiplication was repeated addition, and so $2 \cdot 3$ told us to add 3 2's together. This idea easily covers $0 \cdot 3$. It tells us to add 3 0's together. But, using the definition above, $(0 - 0) + 0 = 0 + 0 = 0$, so $0 \cdot 3$ ought to be 0.

There is more of a problem with $3 \cdot 0$ however. If we apply our basic idea, we find ourselves adding together 0 3's. We have no 3's (or anything else) to work with, so we certainly can't get a *number* for an answer. Our original motivation isn't sufficient to cover this case.

Maybe a natural extension of our original ideas will work. We don't know how $3 \cdot 0$ ought to behave by itself, but maybe we can decide how it should behave when other numbers are around. Let us first consider the expression $4 + 3 \cdot 2$. This seems to tell us to add 2 3's to 4. Next, consider $4 + 3 \cdot 0$. Reasonably, this tells us to add 0 3's to 4, that is, leave 4 alone. The result, of course, is 4. Then, adding $3 \cdot 0$ to 4 is the same as not adding any amount to 4, or by the definition above, it is the same as $4 + 0$. That is, in additions, $3 \cdot 0$ behaves like 0. Since it does so here, we simply take $3 \cdot 0$ to be 0.

Definition 3.2.4 Let x be any whole number. We take $x \cdot 0 = 0 \cdot x = 0$.

Finally exponentiation. The idea we followed was that exponentiation was repeated multiplication, and so 2^3 told us to multiply 3 2's together. This also immediately extends to cover 0^3. We are told to multiply 3 0's together. But by the definition above, $(0 \cdot 0) \cdot 0 = 0 \cdot 0 = 0$. Then 0^3 should be 0. More generally, 0^n should be 0 where n is a counting number.

The case of 3^0 is a problem much like that of $3 \cdot 0$ above. It tells us to multiply 0 3's together. We get no result since we are given no numbers to work with. Again, as above, we may be able to see how 3^0 ought to function as part of a larger calculation. Consider first $4 \cdot 3^2$. This tells us to multiply 4 by 2 3's. Next, consider $4 \cdot 3^0$. This, by analogy, says we are to multiply 4 by 0 3's, that is, leave 4 alone. $4 \cdot 3^0$ ought to be just 4. Multiplying 4 by 3^0 is the same as taking 4 itself, that is, $4 \cdot 1$ or one 4. Since 3^0 behaves like 1 in multiplications, we take 3^0 to be 1. More generally, these considerations lead us to set $x^0 = 1$ where x is any counting number.

Definition 3.2.5 Let n be any counting number. We set $0^n = 0$ and $n^0 = 1$.

Notice that since 0 is not a counting number, 0^0 has not been defined. It is, in fact, a troublesome case. If we extend the clause $0^n = 0$ to allow n to be 0, we are led to $0^0 = 0$. Likewise if we extend the clause $n^0 = 1$ to allow n to be 0, we are led to $0^0 = 1$. Should 0^0 be 0 or 1 then?

In Chapter Two we proved several results about exponentiation in the counting number system. We might try to extend those to the whole number system using the definition above together with:

Possibility 1 $0^0 = 0$;

Possibility 2 $0^0 = 1$.

In this way we could perhaps see which possibility gives rise to a mathematically simpler theory, and use this as a guide in assigning a value to 0^0. But if we try this we quickly discover that every theorem on exponentiation extends to the whole number system equally well under *either* of the possibilities.

Since we have no guide to follow we are best off leaving 0^0 *undefined*. As a matter of fact, in more advanced areas of mathematics it is apparent that *any* choice of a value for 0^0 will have undesirable consequences. So leaving 0^0 undefined is the common course.

From now on, whenever we write x^y it is understood that x and y are whole numbers *not both 0*. We will not explicitly mention this qualification every time an exponentiation appears.

3.3   Basic properties

All the fundamental laws about adding, multiplying and raising to powers for the counting numbers easily extend to include 0. The proofs are quite simple. In Chapter Two we had to show things for infinitely many numbers; now we only have to check one more case. We do a few by way of example, and leave the rest as exercises.

Theorem 3.3.1 *Let x, y and z be whole numbers. Then:*

1. $x + y = y + x$;

2. $x + (y + z) = (x + y) + z$;

3. $x \cdot y = y \cdot x$;

4. $x \cdot (y \cdot z) = (x \cdot y) \cdot z$;

5. $x \cdot (y + z) = x \cdot y + x \cdot z$.

Proof (of part 2.) x, y and z are whole numbers, so one of the following eight situations must hold:

1. x, y and z are all counting numbers;

2. x and y are counting numbers, z is 0;

3. x and z are counting numbers, y is 0;

4. y and z are counting numbers, x is 0;

5. x and y are 0, z is a counting number;

6. x and z are 0, y is a counting number;

7. y and z are 0, x is a counting number;

8. x, y and z are 0.

Please note the system we followed in listing the cases, to make sure we got them all. Now, all we have to to is check that $x + (y + z) = (x + y) + z$ is true in all 8 cases.

In case 1, $x + (y + z) = (x + y) + z$ is true since we proved it in Chapter Two. In case 2, the statement becomes $x + (y + 0) = (x + y) + 0$. But $x + (y + 0) = x + y$ and $(x + y) + 0 = x + y$ by definition, hence in case 2 statement is true.

The other six cases are similar. $\square$

Theorem 3.3.2 *Let x, y and z be whole numbers. Then:*

1. $1^x = 1$;

2. $(x \cdot y)^z = x^z \cdot y^z$;

3. $x^{y+z} = x^y \cdot x^z$;

4. $(x^y)^z = x^{y \cdot z}$.

Exercises

Exercise 3.3.1 Check the other parts of Theorem 3.3.1.

Exercise 3.3.2 Prove Theorem 3.3.2.

Exercise 3.3.3 Show the following continue to hold in the whole number system:

1. $x + 1 = x^+$;

2. $x + y^+ = (x + y)^+$;

3. $x \cdot 1 = x$;

4. $x \cdot y^+ = x \cdot y + x$;

5. $x^1 = x$;

6. $x^{y^+} = x^y \cdot x$.

3.4   Order of whole numbers

Recall our discussion of 'bigger than' for counting numbers. $x > y$ meant, intuitively, we could start counting at y and eventually we would reach x, that is, $x = y^{++\cdots+}$, for some number n of successor symbols. Clearly we don't want n to be 0; this would say we reach x by starting at y and not counting at all, that is, y and x are the same. We really want some successor symbols to be present.

Definition 3.4.1 Let x and y be whole numbers. We say x is *bigger than* y if $x = y + n$ for some *counting number* n.

If x is bigger than y, we write $x > y$. We use the symbols $\geq$, $<$ and $\leq$ as usual.

Theorem 3.4.2 *Let x and y be whole numbers. Then $x > y$ if and only if one of the following is true: 1) both x and y are counting numbers and $x > y$ as defined in Chapter Two, or 2) x is a counting number but y is 0.*

Proof

Part I) Suppose $x > y$. Then $x = y + n$ for some counting number n. We have two possibilities.

a) $y = 0$. Then $x = y + n = 0 + n = n$, so x is a counting number, and item 2) of the theorem holds.

b) $y \neq 0$. Then y must be a counting number. But so is n, hence so is x since it is $y + n$. Then item 1) of the theorem holds.

Part II) Suppose one or the other of items 1) and 2) hold. We have two possibilities.

a) Item 1) holds. Then, by the definition of Chapter Two, $x = y + n$ for some counting number n. Of course this means $x > y$ by the definition of this section.

b) Item 2) holds. Then x itself is a counting number. Now $x = 0 + x$, so $x > 0$ by our present definition, that is, $x > y$. □

Remark By this theorem, we can't have $0 > 0$, or more generally, $0 > x$.

Theorem 3.4.3 *Let x and y be whole numbers. Then $x \geq y$ if and only if $x = y + z$ for some whole number z.*

Theorem 3.4.4 (transitivity) *For any whole numbers x, y and z, if $x > y$, and $y > z$ then $x > z$.*

Theorem 3.4.5 (trichotomy law) *For any whole numbers x and y exactly one of the following holds: $x > y$, $x = y$, $x < y$.*

Proof Again we have a proof by cases. There are four to consider.

1. x and y are counting numbers,

2. x is a counting number, y is 0,

3. x is 0, y is a counting number,

4. x and y are 0.

We take them in order.

Case 1) x and y are counting numbers. Then the result is exactly the statement of the Trichotomy law for counting numbers, as proved in Chapter Two.

Case 2) x is a counting number, y is 0. By Theorem 3.4.2, $x > y$, so one of the three possibilities holds. We can't have $y > x$ because it says $0 > x$ which Theorem 3.4.2 rules out. Also we can't have $x = y$; since x is a counting number but y isn't. Thus in case 2 exactly one of the three possibilities holds.

Case 3) x is 0, y is a counting number. This is similar to case 2.

Case 4) x and y are 0. Then of course $x = y$. By Theorem 3.4.2 we don't have $0 > 0$, so both of $x > y$ and $y > x$ are out. Again exactly one of the three possibilities holds. □

Exercises

Exercise 3.4.1 Show that for any whole numbers x and y, if $x > y$ then $x \geq y^+$.

Exercise 3.4.2 Prove Theorem 3.4.3.

Exercise 3.4.3 Prove Theorem 3.4.4 (Hint: use the proof of transitivity of $>$ for counting numbers).

3.5 Insertion and cancellation

The results of Chapter Two extend easily to include 0. We state the extensions, and leave the verifications as exercises.

Theorem 3.5.1 *For any whole numbers x, y and z, $x = y$ if and only if $x + z = y + z$.*

Theorem 3.5.2 *For any whole numbers x, y and z, $x > y$ if and only if $x + z > y + z$.*

Theorem 3.5.3 *For any whole numbers x, y and z with z not 0, $x > y$ if and only if $x \cdot z > y \cdot z$.*

Even though z can't be 0 in the theorem above, we have the following:

Theorem 3.5.4 *For any whole numbers x, y and z, if $x \geq y$ then $x \cdot z \geq y \cdot z$.*

Theorem 3.5.5 *For any whole numbers x, y and z, with z not 0, $x = y$ if and only if $x \cdot z = y \cdot z$.*

Theorem 3.5.6 *For any whole numbers x, y and z with $z \neq 0$, $x > y$ if and only if $x^z > y^z$.*

Theorem 3.5.7 *For any whole numbers x, y and z, with z not 0 or 1, $x > y$ if and only if $z^x > z^y$.*

Theorem 3.5.8 *For any whole numbers x, y and z,*

1. *if $z \neq 0$, then $x = y$ if and only if $x^z = y^z$,*

2. *if $z \neq 0$, $z \neq 1$, then $x = y$ if and only if $z^x = z^y$.*

Exercises

Exercise 3.5.1 Prove Theorem 3.5.1.

Exercise 3.5.2 Prove Theorem 3.5.2.

Exercise 3.5.3 Prove that for any whole numbers x and y, if $x^+ > y$ then $x \geq y$.

Exercise 3.5.4

1. Prove Theorem 3.5.3.

2. Show the result is false if the restriction $z \neq 0$ is dropped.

Exercise 3.5.5 Prove Theorem 3.5.4.

Exercise 3.5.6

1. Prove Theorem 3.5.5.

2. Show the clause $z \neq 0$ is necessary.

Exercise 3.5.7

1. Prove Theorem 3.5.6.

2. Why is the restriction, $z \neq 0$, necessary?

Exercise 3.5.8

1. Prove Theorem 3.5.7.

2. Account for the restriction on z.

Exercise 3.5.9

1. Prove Theorem 3.5.8.

2. Account for the restrictions on z.

3.6 Subtraction

The idea behind addition was that it was counting, not necessarily starting at 1. Thus, $7 + 4$ tells us to start at 7 and count off the next 4 numbers. It would be useful to have available a similar operation, but involving counting backwards. We call this *subtraction*, and symbolize it by $x - y$. We may think of $7 - 4$ as telling us to start at 7 and count backwards 4 numbers. But as might be expected, for our official definition, we would rather find an alternative characterization that won't require us to count things.

Using our intuitive ideas, what is $7 - 4$? There is an easy way of seeing this. Using our definition of addition we can easily compute that $3 + 4 = 7$. But $3 + 4$, informally tells us to start at 3 and count off the next 4 numbers. Then, if starting at 3 and counting forward 4 produces 7, certainly starting at 7 and counting backwards 4 should produce 3. In effect, we can verify that $7 - 4$ is 3 by showing that $3 + 4$ is 7. This suggests the following.

Tentative Definition $a - b$ is that number c for which $c + b = a$.

There are two problems with this. First, for any particular a and b, how do we know when there will be such a number c? Second, if there is an appropriate number c, how do we know there is only one?

The first problem is easy to deal with. We simply announce that $a - b$ is to make sense only when there is a number c for which $c + b = a$. But, by Theorem 3.4.3, this happens precisely when $a \geq b$. Thus, we only define $a - b$ when $a \geq b$.

As to the second problem, suppose we had more than one candidate for c. That is, suppose we had whole numbers c and c' for which $c + b = a$ and also $c' + b = a$. Then certainly $c + b = c' + b$ and by Theorem 3.5.1, $c = c'$. So, in fact, only one such number can exist.

Definition 3.6.1 Let x and y be any whole numbers. If $x \geq y$ we say $x - y$ is defined, and it is that number z such that $z + y = x$.

According to this definition, $x - y$ is that z such that $z + y = x$. That is, $x - y$, when added to y, gives x. Also, according to this definition, $x - y$ is that unique z such that $z + y = x$. So if $z + y = x$, z must be $x - y$. We list these observations in the following.

Theorem 3.6.2 *Let x and y be whole numbers with $x \geq y$.*

1. *$(x - y) + y = x$,*

2. *if $z + y = x$ then $z = x - y$.*

Exercises

Exercise 3.6.1 Show $9 - 3$ is defined and $9 - 3 = 6$.

Exercise 3.6.2 Show $x - x$ is defined and $x - x = 0$.

Exercise 3.6.3 Show $x - 0$ is defined and $x - 0 = x$.

Exercise 3.6.4 Show $x^+ - x$ is defined and $x^+ - x = 1$.

3.7 Basic properties of subtraction

The basic properties of subtraction have not been as intensively studied as have those of addition and multiplication. We have no names comparable to associativity and commutativity, for example. The reason for this is simply that once negative numbers are introduced, the properties of subtraction become properties of addition and so do not need special treatment. However, the basic facts about subtraction were known and used long before the concept of negative number was developed. Moreover, the properties of subtraction were part of the original motivation for creating negative numbers. For these reasons and others we develop properties of subtraction in some detail. We begin with some relationships with order.

Theorem 3.7.1 *Let x, y and z be counting numbers, with $x \geq z$ and $y \geq z$. Then $x > y$ if and only if $x - z > y - z$.*

Proof By Theorem 3.5.2, $x - z > y - z$ is equivalent to $(x - z) + z > (y - z) + z$. But by Theorem 3.6.2, this is equivalent to $x > y$. □

Theorem 3.7.2 *Let x, y and z be counting numbers, with $x \geq y$ and $x \geq z$. Then $y > z$ if and only if $x - y < x - z$.*

Proof By Theorem 3.5.2, $x - y < x - z$ is equivalent to $(x - y) + (y + z) < (x - z) + (y + z)$. This in turn is equivalent to $[(x - y) + y] + z < [(x - z) + z] + y$ And by Theorem 3.6.2, this is equivalent to $x + z < x + y$. Finally, by Theorem 3.5.2 again, this is equivalent to $z < y$. □

Most of the proofs which follow have the same general pattern. We give this pattern here in schematic form.

Suppose we have two expressions, say A and B, which we want to prove equal. Say we can find some clever thing, C, which when added to each produces obviously the same result. That is, $A + C$ and $B + C$ are recognizably identical. Then:

$$A + C = B + C$$

so by the cancellation law for addition,

$$A = B.$$

In short, if adding the same thing to both A and B produces the same result, A and B must have been the same to begin with.

Theorem 3.7.3 *Let x, y and z be whole numbers with $x \geq z$ and $z \geq y$. Then $x - y = (x - z) + (z - y)$.*

Remark $x \geq z$ and $z \geq y$ so both subtractions on the right hand side are defined. Also, by transitivity, $x \geq y$, so the subtraction on the left hand side is defined.

Proof Let us add y to each of the expressions $x - y$ and $(x - z) + (z - y)$.

$$
\begin{aligned}
(x - y) + y &= x && \text{(Theorem 3.6.2)} \\
\text{Also,} \\
[(x - z) + (z - y)] + y &= (x - z) + [(z - y) + y] && \text{(associativity)} \\
&= (x - z) + z && \text{(Theorem 3.6.2)} \\
&= x && \text{(Theorem 3.6.2)}
\end{aligned}
$$

Since we get the same results in both cases, $x - y$ and $(x - z) + (z - y)$ must have been the same to begin with (recall, the cancellation law for addition is being used here). $\square$

Theorem 3.7.4 (Distributive law for subtraction)
Let x, y and z be whole numbers with $y \geq z$. Then, $x \cdot (y - z) = x \cdot y - x \cdot z$.

Remark $y \geq z$, so $y - z$ is defined. Also, since $y \geq z$, $x \cdot y \geq x \cdot z$, so $x \cdot y - x \cdot z$ is defined.

Proof We add $x \cdot z$ to the two expressions in question.

$$
\begin{aligned}
(x \cdot y - x \cdot z) + x \cdot z &= x \cdot y && \text{(Theorem 3.6.2)} \\
\text{and} \\
x \cdot (y - z) + x \cdot z &= x \cdot [(y - z) + z] && \text{(distributive law)} \\
&= x \cdot y && \text{(Theorem 3.6.2)}
\end{aligned}
$$

This establishes the theorem. $\square$

Theorem 3.7.5 *Let x, y and z be whole numbers with $y \geq z$. Then $(x + y) - z = x + (y - z)$.*

Theorem 3.7.6 *Let x, y and z be whole numbers with $x \geq y + z$. Then $x - (y + z) = (x - y) - z = (x - z) - y$.*

Theorem 3.7.7 *Let x, y and z be whole numbers with $x \geq y$. Then $x - y = (x + z) - (y + z)$.*

Theorem 3.7.8 *Let x, y and z be whole numbers. Then:*

1. *if $x + z \geq y \geq z$, $x - (y - z) = (x + z) - y$;*

2. *if $x \geq y \geq z$, $x - (y - z) = (x - y) + z$.*

Exercises

Exercise 3.7.1 Let x, y, z and w be whole numbers and suppose $x > y$ and $z < w$. Show $x - z > y - w$.

Exercise 3.7.2 Let x, y, z and w be whole numbers and suppose $x \geq y$ and $z \geq w$. Show $x \cdot z + y \cdot w \geq x \cdot w + y \cdot z$.

Exercise 3.7.3 Show all subtractions in Theorem 3.7.5 are defined and establish the theorem.

Exercise 3.7.4 Show all subtractions in Theorem 3.7.6 are defined, and establish the theorem.

Exercise 3.7.5 Show all subtractions in Theorem 3.7.7 are defined, and establish the theorem.

Exercise 3.7.6 Show all subtractions in Theorem 3.7.8 are defined, and establish the theorem. Hint: For part 1 Theorem 3.7.3 is useful. Part 2 can be derived from part 1, or proved directly.

Exercise 3.7.7 Let x, y, z and w be whole numbers with $x \geq y$ and $z \geq w$. (See Exercise 3.7.2.) Show $(x - y) \cdot (z - w) = (x \cdot z + y \cdot w) - (x \cdot w + y \cdot z)$.

3.8 Archimedean order

> ...the greater exceeds the less by such a magnitude as, when added to itself, can be made to exceed any assigned magnitude among those which are comparable with one another.
>
> *On the Sphere and Cylinder,*
> Book I
> –Archimedes, c. 287 – c. 212 BCE

Even though we are working with the whole number system, we are about to prove a result about the counting numbers. We want to show that by adding enough y's together, we can make the result as large as we want. Now, no matter how many 0's we add, the result is still 0, so we restrict y to be a counting number. Then it seems plausible that by adding enough y's together we can make the result arbitrarily large. A mathematical structure in which this happens is *Archimedean ordered*.

According to the ideas which motivated our definition of multiplication, adding k y's together should be $k \cdot y$. So, in effect, to say we have Archimedean order is to say $k \cdot y$ can be made as large as desired by taking k suitably big. More precisely, if $y \leq x$, no matter how big x is, by taking k suitably large, $k \cdot y$ will be bigger than x.

What we will actually prove here is a more precise version of this: not only will $k \cdot y$ be bigger than x for some k, but there is a smallest k which will do.

More precisely yet, we will show there is some q such that $q^+ \cdot y$ is bigger than x though $q \cdot y$ is not. In fact, the q is *uniquely* determined by x and y.

Lemma 3.8.1 *Let x and y be counting numbers. $x^+ \cdot y > x$.*

Proof We know $x^+ > x$. Also, since y is a counting number, $y \geq 1$. Then $x^+ \cdot y \geq x^+ \cdot 1 = x^+ > x$. $\square$

Theorem 3.8.2 *Let x and y be counting numbers, with $y \leq x$. There is a counting number q such that $q^+ \cdot y > x$ but $q \cdot y \leq x$.*

Proof Let S be the following collection of counting numbers. Put n in S if $n \cdot y > x$. S is not empty since, by Lemma 3.8.1, $x^+ \in S$. Then by the well-ordering principle, Theorem 2.13.1, S has a smallest member, call it k. Now $1 \cdot y = y \leq x$, so $1 \notin S$. But $k \in S$, so $k \neq 1$. Then k has a predecessor, say $k = q^+$. Then q^+ is the smallest member of S.

Since $q^+ \in S$, $q^+ \cdot y > x$. Since $q < q^+$ and q^+ is the smallest member of S, $q \notin S$. Then we don't have $q \cdot y > x$, so $q \cdot y \leq x$. $\square$

Exercises

Exercise 3.8.1 Show there is only one number q meeting the conditions of Theorem 3.8.2. Hint: suppose there were two, say q and Q. Then:

$$q^+ \cdot y \;>\; x \qquad\qquad Q^+ \cdot y \;>\; x$$
$$q \cdot y \;\leq\; x \qquad\qquad Q \cdot y \;\leq\; x.$$

Now show contradictions follow from either of $q < Q$ or $q > Q$.

3.9 Exact division

Just as we defined subtraction to be opposite to addition, we want to define an operation of division to be opposite to multiplication. Recall, our definition of subtraction was:

$$x - y \text{ is that unique } z \text{ such that } z + y = x.$$

In the same fashion, we may try

$$x \div y \text{ is that unique } z \text{ such that } z \cdot y = x.$$

The problem is, when will $x \div y$ make sense? We know $x - y$ makes sense whenever $x \geq y$. There is nothing quite as simple available for division, so we simply say $x \div y$ makes sense whenever it makes sense.

Definition 3.9.1 Let x and y be whole numbers. We say x is *divisible by y* if there is one and only one z for which $z \cdot y = x$. If there is such a z, $x \div y$ is that z.

We consider some examples. First, is 6 divisible by 2? Well, $z \cdot 2 = 6$ when $z = 3$. If no other value of z will work, then 6 is divisible by 2. But we have the following general fact.

Lemma 3.9.2 *Suppose $z \cdot y = x$ and $z' \cdot y = x$ where y is not 0. Then $z = z'$.*

This Lemma tells us, if we are not trying to divide by 0, and we have one candidate for $x \div y$, it is in fact the only one so $x \div y$ is defined. Now, continuing our example, $3 \cdot 2 = 6$, and by the Lemma, 3 is the only value of z for which $z \cdot 2 = 6$. Then 6 is divisible by 2, and $6 \div 2 = 3$.

Next, is 6 divisible by 0? Is there some z for which $z \cdot 0 = 6$? No, since $z \cdot 0 = 0$ no matter what z is. 6 is *not* divisible by 0.

More generally, no matter what x is, x is not divisible by 0.

Finally, is 5 divisible by 2? No, which we may show as follows. By simple calculations, we have:

$$0 \cdot 2 = 0, \text{ not } 5$$
$$1 \cdot 2 = 2, \text{ not } 5$$
$$2 \cdot 2 = 4, \text{ not } 5$$
$$3 \cdot 2 = 6, \text{ not } 5.$$

And if $z > 3$, $z \cdot 2 > 3 \cdot 2$ so $z \cdot 2$ is not 5. Thus, no matter what z is, $z \cdot 2$ is never 5, so 5 is not divisible by 2.

Exercises

Exercise 3.9.1 Prove Lemma 3.9.2.

Exercise 3.9.2 Show 0 is not divisible by 0.

Exercise 3.9.3 Establish which of the following are defined, and for those which are, produce a value:

1. $8 \div 4$;
2. $9 \div 4$;
3. $0 \div 5$;
4. $x \div 1$.

3.10 Division with remainder

Suppose we try to see if 7 is divisible by 3. Well,

$$0 \cdot 3 = 0 \text{ not } 7$$
$$1 \cdot 3 = 3 \text{ not } 7$$
$$2 \cdot 3 = 6 \text{ not } 7$$
$$3 \cdot 3 = 9 \text{ not } 7$$

and from here on, things will be too big. 7 is not divisible by 3.

Suppose we revise the question. Even though we can't fit an exact number of 3's into 7, what's the most we can fit in? A glance at the list above shows 2 3's will fit but 3 3's won't. The 2 3's don't exactly make up 7, they fall short by 1. Then $2 \cdot 3 + 1 = 7$. The standard terms here are: 2 is the *quotient*, 1 is the *remainder* in our attempt to divide 7 by 3.

More generally, suppose we try to divide x by y, and suppose it turns out we can't. Well, let q be the largest number of y's which will fit into x, and let r be the amount by which $q \cdot y$ falls short of x. Then

$$q \cdot y + r = x. \tag{3.1}$$

Here q is the *quotient* and r is the *remainder*. A simple observation: the remainder, r, must be smaller than y, since otherwise we could have fit more y's into x. That is,

$$r < y. \tag{3.2}$$

It turns out that even when x is not divisible by y, there must still exist a quotient q and a remainder r satisfying (3.1) and (3.2). And, moreover, they are *unique*. There are some qualifications to be made, however. First, we don't want y to be 0. There are many reasons for this, an obvious one being that if y is 0, no r can satisfy (3.2). Second, we want to allow the possibility of a 0 remainder, which means we had exact division after all. That way we don't have to treat exact and non-exact division as separate cases.

Now we state and prove a formal version of the above.

Theorem 3.10.1 *Let x and y be whole numbers, with $y \neq 0$. There are unique whole numbers q and r satisfying the conditions:*

1. $x = q \cdot y + r$;

2. $r < y$.

Proof There are two parts to the proof, the existence of q and r, and their uniqueness. We establish uniqueness first.

I. Suppose we had two quotients, q and Q, and two remainders, r and R. We show that, in fact, they are the same. Now q and r satisfy (3.1) and (3.2), as do Q and R. Hence:

$$\begin{aligned} x &= q \cdot y + r \\ r &< y \end{aligned}$$

and also

$$\begin{aligned} x &= Q \cdot y + R \\ R &< y. \end{aligned}$$

Now, $x = q \cdot y + r \geq q \cdot y$ (Theorem 3.4.3) so $x \geq q \cdot y$. Similarly $x \geq Q \cdot y$. Also $x = q \cdot y + r < q \cdot y + y = q^+ \cdot y$ so $x < q^+ \cdot y$. Similarly $x < Q^+ \cdot y$. Then Exercise 3.8.1 says $q = Q$. Finally, $x = q \cdot y + r$ and $x = Q \cdot y + R$, so

$q \cdot y + r = Q \cdot y + R$. But $q = Q$, so $q \cdot y + r = q \cdot y + R$. Then by cancellation, $r = R$.

This completes the proof of uniqueness.

II. Now we show q and r must exist. We consider two cases.

case 1) $x < y$,

case 2) $x \geq y$.

The first case is not very interesting; we are dividing y into a smaller number. So we go directly to case 2).

Case 2) $x \geq y$. Now, $y \neq 0$, so y is a counting number. Since $x \geq y$, x is a counting number too. Then we may apply Theorem 3.8.2 on Archimedean order. There is a counting number q such that $q^+ \cdot y > x$ but $q \cdot y \leq x$. Since $x \geq q \cdot y$, $x - q \cdot y$ is defined. Call it r. That is, $x - q \cdot y = r$, from which follows $x = q \cdot y + r$. We must yet show $r < y$. Well, $x < q^+ \cdot y$ but $x = q \cdot y + r$, so $q \cdot y + r < q^+ \cdot y = q \cdot y + y$ so by cancellation, $r < y$. In case 2) we have shown the existence of appropriate numbers q and r. $\square$

Exercises

Exercise 3.10.1 Complete the proof of Theorem 3.10.1 by showing in case 1) suitable choices for q and r exist.

Exercise 3.10.2 Use Theorem 3.10.1 to prove that 9 is not divisible by 4.

CHAPTER 4

Naming Numbers — Place Value Notation

"I don't *rejoice* in insects at all," Alice explained, "because I'm rather afraid of them – at least the large kinds. But I can tell you the names of some of them."

"Of course they answer to their names?" the Gnat remarked carelessly.

"I never knew them to do it."

"What's the use of their having names," the Gnat said, "if they won'n't answer to them?"

"No use to *them*," said Alice; "but it's useful to the people that name them, I suppose. If not, why do things have names at all?"

"I ca'n't say," the Gnat replied.

> *Through the Looking Glass*,
> Chapter Three,
> Lewis Carroll, 1871

4.1 Introduction

There is an unlimited supply of whole numbers. People don't use them all in their everyday affairs, but they do use so many that some *system* of referring to them is a necessity. Over the course of human history many systems for naming numbers have been tried. This is not the place to review them. Today, however, virtually the whole world has settled on a *place-value* system of names, using a *base* of 10. But other bases are also in use for various special purposes. Computers do their work in base 2; people who work with computers commonly use base 16, and sometimes base 8; and until recently there was a group who advocated a switch in the common system from base 10 to base 12.

Now the particular choice of a base, while having a practical effect, makes no difference in theory. So we present place-value naming systems for all bases. Partly this will allow you to appreciate the key features of the conventional system, by seeing how these features work in unfamiliar settings. Then, too, having many bases around provides us with a source of exercises. Finally, as we noted above, bases other than 10 are in use today.

In a base n system one introduces special *symbols*, called *digits*, to denote the first n whole numbers (from 0 to $n - 1$). Then one uses *finite strings* or *sequences* of these digits to name all numbers, according to a certain convention. Then right at the start we are faced with two problems: what is a digit, and what is a finite sequence. We take up these issues in order.

It is common, in educational circles, to distinguish between *number* and *numeral*. A numeral is a *symbol* that we use to *name* a number. Thus the symbol '1', written as a vertical stroke, is used to name the first counting number. It is the symbol that occurs in our discourse, not the number itself. Just as when we say, "George has red hair," we use a *name* for George; we do not use George himself.

But if we try to follow up on this carefully, we find some difficulties. There is only one symbol we ordinarily use to name the first counting number, the symbol '1'. But '1' may occur many times on a page. Since no single occurrence is more fundamental than any other, which is *the* symbol? The usual response is that there is one symbol, with many instances, and the instances of the symbol are what we see on the page. But then, what is the *symbol itself* that has all these instances? We are rapidly led to the position that there is some class of mental concepts called *symbols*, but a symbol is something different than any of the instances of it. In short, symbols become abstractions much like numbers themselves.

Clearly there are difficulties here, and they go rather deep. Unless we want this to become a treatise on philosophy, we had better avoid the question of "what is a place-value name," which first requires an answer to the question, "what is a symbol." We should concentrate on the question, "How does a system of place-value names *work*?"

To handle this question, we don't need to know what digits *are*, in any philosophical sense. We only need to have some things that we can *call* digits, and we can go on from there. We do the simplest thing, and choose our digits from among the numbers themselves. Thus the base 10 digits, for us, are the *numbers* 0, 1, 2, 3, 4, 5, 6, 7, 8, 9, and a base 10 name is a finite sequence of these digits. This lets us carry out our investigation without getting involved in endless complications.

Thus we have actually side-stepped the whole philosophical issue of what symbols are in the interests of getting on with the mathematical side of things.

A place-value name is a finite sequence or string of digits. Thus if we write something like 216, most people understand we have in mind the finite sequence consisting of the three digits, 2, 1 and 6, in that order. But this means that, for this chapter, we need a *new* mathematical object, a *finite sequence*. Now, it is possible to define a notion of finite sequence, given the usual machinery of set theory. Indeed, there are several ways of doing so, many of them quite natural. It is even possible to define something meeting all the customary technical requirements of a finite sequence, using just the machinery of arithmetic that has been developed thus far, though most people would say this is not what they *mean* by finite sequence, even if it does the job technically. But still, if we have all these alternatives, most of which bring in substantial set-theoretic equipment, what 'real' meaning are we to assign to the notion?

Well, all the alternative ways of defining a notion of finite sequence have certain features in common, and it is really these features that we make use of in practice. For example, every finite sequence has a *length*; that length is a counting number; and so on. Whatever it is that people 'mean' by finite sequence, they all agree on such things. So, we have decided not to choose any of the many candidates for a definition of finite sequence as fundamental. Rather, we will simply announce there are such things as finite sequences, and their behavior meets certain conditions which we will specify. In short, we treat the notion of finite sequence *axiomatically*, just as we did the notion of counting number itself.

So, for this chapter, our assumed mathematical machinery is enlarged. We have numbers, and we have finite sequences. We do not say what either *are*, but we do say what we expect of their *behavior*.

4.2   Place–value names

> Moses chose able men out of all Israel, and made them heads over the people, rulers of thousands, of hundreds, of fifties, and of tens.
>
> –The Bible
> Exodus 18:25

We introduce a system of base n names for whole numbers, for each $n \geq 2$, and we say how these names are to be thought of as naming numbers. Though any $n \geq 2$ is possible, our examples will generally involve bases 10, the conventional one, as well as bases 2, 5 and 12.

The preceding paragraph exemplifies a certain difficulty in writing a chapter like this one. We are about to introduce place-value notation, yet we just used it, when we wrote 10, 2, 5 and 12 above. You are going to have to keep mentally balanced two levels of discourse. There is the subject matter we are talking *about*, and there is the language we *use* in our discourse about that subject matter. In the *formal* development thus far, in our definitions and theorems and proofs, we have not used place-value notation. This is as it should be; we have not yet formally introduced it or developed its properties. But the language we use for our informal discourse, when we talk *about* our subject, is Common English. And we think of this as including the usual machinery for naming numbers; that is, common base 10 notation.

Our use of base 10 notation 'on the outside' is of no theoretical importance, of course; and we have been doing it all along, probably without it being noticed. For instance, we have been calling our exercises things like 2.1, 2.2, 2.3, etc. Clearly we could have called them Tom, Dick, Harry, etc., just as well, but this would not have been as convenient. However, in this chapter, where place-value notation itself is being developed, a problem could arise unless the reader is careful.

From now on, in our *language of discourse* we will feel free to use standard number names. In our *formal subject matter* however, we will not do this. Rather, we will carefully define place-value naming systems and develop their properties.

Definition 4.2.1 Let n be a whole number with $n \geq 2$. The base n digits are those whole numbers $< n$.

Example The base 2 digits are 0 and 1. The base 5 digits are 0, 1, 2, 3, 4. The base 10 digits are 0, 1, 2, 3, 4, 5, 6, 7, 8, 9. For base 12 we need the base 10 digits together with two more. For this purpose we define $t = 9^+$ and $e = t^+$.

We said the base 2 digits were 0 and 1. This really should be proved. But it is rather easy, given all the work of earlier chapters.

First, $0 + 2 = 2$, hence $0 < 2$ by the definition of $<$ in Chapter Three §4. Thus 0 is a base 2 digit. Likewise $1 + 1 = 1^+ = 2$, so $1 < 2$. This means 1 is a base 2 digit.

On the other hand, suppose x is a base 2 digit, but $x \neq 0$ and $x \neq 1$. We show this is impossible. Since x is a base 2 digit, $x < 2$. Then $x \leq 1$. Since $x \neq 1$ we have $x < 1$. Repeating, we get $x \leq 0$. Since $x \neq 0$ we have $x < 0$, and this violates the Trichotomy Law 3.4.5. Thus 0 and 1 are the only base 2 digits.

In a similar way our assertions about the various other bases can be verified.

From now on, any reference to *base n* presupposes it makes sense, namely that n is a counting number with $n \geq 2$. We will not say all this each time.

We introduce base n names: finite sequences of base n digits. For the reasons given in the previous section, we do so axiomatically. Thus we assume there is a collection of objects called *base n names*. We assume each base n name has a *length* which is a counting number. And we assume there is an operation of *concatenation*, intuitively the following of one name by another to produce a longer name. If w and z are base n names, we denote the result of concatenating them (in the order given) by wz. For example, 216 and 38 are base 10 names (as we can show in a moment). Concatenating them we get 21638.

Now, we assume all this machinery meets the following conditions, or *name axioms*.

N1 Concatenation is associative (and thus we will not need parentheses to indicate grouping).

N2 Every base n digit d is also a base n name of length 1.

N3 If w is a base n name of length k and d is a base n digit then wd is a base n name of length k^+.

N4 If z is a base n name of length 1 then z is some base n digit, d.

N5 If z is a base n name of length k^+ then there is a unique base n name w of length k and there is a unique base n digit d such that $z = wd$.

Example We show 231 is a base 5 name of length 3.

a) 2 is a base 5 digit, hence 2 is a base 5 name of length 1 by *N2*.

b) Since 2 is a base 5 name of length 1, and 3 is a base 5 digit, 23 is a base 5 name of length $1^+ = 2$ by *N3*.

c) Since 23 is a base 5 name of length 2, and 1 is a base 5 digit, 231 is a base 5 name of length $2^+ = 3$ by *N3* again.

Note that 231 is also a base 10 name. Indeed, it is a name in base n for any $n \geq 4$.

Now we introduce some useful terminology, that of *last term* and *first term*.

Definition 4.2.2 Let z be a base n name. We define the last term of z as follows. The length of z is a counting number, hence is either 1 or is k^+ for some k. If the length of z is 1 then by N4, z is some base n digit, d; in this case, the last term of z is d. If the length of z is k^+, then by N5, $z = wd$ for some unique name w and digit d; in this case the last term of z is d.

Example 231 is a base 5 name of length 3, so case 2 above applies. 231 is 23 concatenated with the digit 1, hence the last term of 231 is 1.

Definition 4.2.3 Let z be a base n name. We define the first term of z as follows (again using the same two cases as in the previous definition). If the length of z is 1 then z is some base n digit d; in this case the first term of z is d. If the length of z is k^+ then $z = wd$ for some unique name w of length k and digit d. In this case the first term of z is whatever the first term of w is.

Notice that this will not generally tell us outright what the first term of a name is, but it will let us calculate it with a little work.

Example 231 is a base 5 name of length 3, so its first term is whatever the first term of 23 is. Then, 23 is a name of length 2, so its first term is whatever the first term of 2 is. Finally, 2 is a name of length 1, so its first term is 2. Thus the first term of 231 is 2.

Definition 4.2.4 We call a base n name *proper* if either it is of length 1, or it is of length > 1 but its first term is not 0.

Example 231 is proper. 0231 is not proper. 0 is proper.

Now we say how our names are to be thought of as naming numbers. This, of course, is what we are really interested in. We need to be able to distinguish between the name and the number being named. The notational convention we follow is this. 217, say, can be thought of as a base 10 *name*. As such, it is a finite sequence of digits of length 3 and is not, itself, a number. But it can be thought of as *naming* a number, as we describe below, in the base 10 system. We use $(217)_{10}$ for that number. Thus 217 is, for us, a finite sequence that names the number $(217)_{10}$ in base 10 notation. Now for the details.

The idea is that in base 10, say, each shift to the left by one place means values jump by a factor of 10. Likewise, in base 2, each shift left one place should multiply values by a factor of 2. Hence the term *place-value*, since the value of a digit appearing in a name depends not only on the digit itself, but also on how far to the left it occurs. Now in our formal definition we only need to say all this for a *single* shift to the left, since a shift to the left of many places is the result of many single shifts.

Definition 4.2.5 Let z be a base n name. We define $(z)_n$ using a familiar two-case arrangement as follows. If the length z is 1, then z is a single base n digit, d and we take $(z)_n$ to be the number d. If the length of z is k^+, then $z = wd$ for some unique name w of length k, and digit d; then $(z)_n = (wd)_n = (w)_n \cdot n + d$.

Note that, as in the definition of first term, this does not say outright what $(z)_n$ is, but rather it allows a value for $(z)_n$ to be calculated.

Example We have already seen that 231 is a base 5 name. Now we see what number it names.

a) Since 2 is a base 5 digit, $(2)_5 = 2$ by clause 1 of the definition.

b) Then $(23)_5 = (2)_5 \cdot 5 + 3$ by clause 2, and by part *a* this is equal to $2 \cdot 5 + 3$.

c) $(231)_5 = (23)_5 \cdot 5 + 1$ by clause 2 again, and by part *b* this is just $(2 \cdot 5 + 2) \cdot 5 + 1$.

Actually we can go on a little with this example, using the distributive law and the properties of exponents (and leaving out some of the parentheses).

$$
\begin{aligned}
(231)_5 &= (2 \cdot 5 + 3) \cdot 5 + 1 \\
&= 2 \cdot 5 \cdot 5 + 3 \cdot 5 + 1 \\
&= 2 \cdot 5^2 + 3 \cdot 5^1 + 1 \cdot 5^0
\end{aligned}
$$

In our everyday place value system the string 10 names the base. Exercise 4.2.3 says that this applies no matter what the choice of base. Likewise Exercise 4.2.4 says that another familiar feature of the customary system also carries over to all bases, namely: multiplying by the number that 10 names corresponds to adjoining another 0 to the name.

In giving examples and exercises, it is often necessary for us to say what base we have in mind. If so, we do this using standard base ten notation. Thus we may write $(216)_{12}$ to indicate we are naming in the base twelve system. This convention should cause no theoretical problems.

If we have many numbers to discuss and all of them are named using base n notation, it will often be convenient to drop the subscripts n, and simply announce, "all these are numbers named in the base n system," or more simply, "we use base n notation throughout." So, if we say something like, "in base 5, $4^+ = 10$", what we mean is $(4)_5^+ = (10)_5$. We use this convention in several of the exercises.

Finally, we turn to the issue of *proper* names, as opposed to what we might call *improper* ones.

Theorem 4.2.6 *Let z be a base n name. Then so is $0z$, and $(0z)_n = (z)_n$.*

Proof We use induction on the length of z. That is, form a collection C of counting numbers as follows. Put k into C provided, for every base n name z of length k, $(0z)_n = (z)_n$. If we show every counting number is in C, it will establish the theorem. Of course we do this by showing C is inductive.

Suppose z is a name of length 1. Then z must be a base n digit, say d. Then, using both parts of the definition given earlier,

$$
\begin{aligned}
(0z)_n &= (0d)_n \\
&= (0)_n \cdot n + d \\
&= 0 \cdot n + d \\
&= 0 + d \\
&= d \\
&= (d)_n \\
&= (z)_n
\end{aligned}
$$

Since z was an arbitrary name of length 1, it follows that $1 \in C$.

Suppose $k \in C$. We show $k^+ \in C$. Well, let z be any base n name of length k^+. Then $z = wd$ where w is a name of length k and d is a digit. Since w is of length k, and we are assuming $k \in C$, we have that $(0w)_n = (w)_n$. But then

$$
\begin{aligned}
(0z)_n &= (0wd)_n \\
&= (0w)_n \cdot n + d \\
&= (w)_n \cdot n + d \\
&= (wd)_n \\
&= (z)_n
\end{aligned}
$$

Since z is an arbitrary name of length k^+, it follows that $k^+ \in C$.

Thus C is inductive, and we are done. $\square$

Exercises

Exercise 4.2.1 Show that for any two base n names z_1 and z_2, the length of the name $z_1 z_2$ is the length of z_1+ the length of z_2. Hint: use induction on the length of z_2. See the proof of Theorem 4.2.6 if trouble arises.

Exercise 4.2.2 Explain why 0 and 1 are ruled out as bases.

Exercise 4.2.3 Show that $(10)_n = n$ for any base n.

Exercise 4.2.4 Let w be a base n name. Then $w0$ is also a base n name. Show that $(w0)_n = (w)_n \cdot (10)_n$.

Exercise 4.2.5 Show the following:

1. $(11)_2 = (3)_{10}$;

2. $(1)_2^+ = (10)_2$;

3. $(10)_2^+ = (11)_2$;

4. $(1)_{10}^+ = (2)_{10}$;

5. $(2)_{10}^+ = (3)_{10}$;

6. $(11)_2 = (3)_{10}$.

Exercise 4.2.6

1. Using base 10 notation throughout, show that $100 = 10^2$ and $1000 = 10^3$;

2. Using base 2 notation through, show that $100 = 10^{10}$ and $1000 = 10^{11}$.

Exercise 4.2.7

1. Using base 10 notation throughout, show that $231 = 2 \cdot 10^2 + 3 \cdot 10^1 + 1 \cdot 10^0$.

2. Using base 2 notation throughout, show that $1111 = 1 \cdot 10^{11} + 1 \cdot 10^{10} + 1 \cdot 10^1 + 1 \cdot 10^0$.

Exercise 4.2.8 Let w be a base n name of length c. Show $(w)_n < n^c$. Hint: use induction on c, and recall that if $x < y$ then $x^+ \leq y$.

Appendix

This is for those of you who read Chapter Two Section 2.5.

The definitions of first term and of $(z)_n$ are *implicit* definitions. As we explained in Chapter Two, such definitions should be shown to be equivalent to *explicit* ones in order to be sure something really is being defined. Now in Chapter Two, our main tool for doing this was Theorem 2.5.1, on Definitions. We can not use that now, because it talked about *numbers*, while now we are interested in *names*. It is possible, however, to prove a similar theorem about names which we could use to 'justify' the implicit definitions earlier in this section. The wording and proof of such a theorem are similar to those of Theorem 2.5.1 and we omit them. We will feel free, however, to make frequent use of implicit definitions in this Chapter.

4.3 Successor

> There are some, king Gelon, who think that the number of the sand
> is infinite in multitude.... Again there are some who, without regarding
> it as infinite, yet think that no number has been named which is great
> enough to exceed its multitude.... But I will try to show you by means
> of geometrical proofs, which you will be able to follow, that, of the num-
> bers named by me...some exceed not only the number of the mass of
> sand equal in magnitude to the earth...but also that of a mass equal in
> magnitude to the universe.

> *–The Sand Reckoner*
> Archimedes

As we have known since childhood, counting is easy using place-value names.
That is, there is a simple *algorithm* or rule for turning a name of x into a name
of x^+.

The rule for going from a base n name to a base n name for the next whole
number may be loosely stated as follows. Look at the right hand digit. If there
exists a bigger digit, replace the right hand digit by the next bigger, then stop.
Otherwise, replace it by 0, move left one place, and repeat the process. We now
want to state this a little more formally. In the definition below, for each base
n name z we define another base n name which we denote z^s, intended to name
the successor to whatever z named. z^s is not to be read as an exponentiation;
s is not a number. Both z and z^s are simply base n names.

Definition 4.3.1

1. If d is a base n digit then $d^s = d^+$ if d is not the biggest base n digit, and
   $d^s = 10$ if d is the biggest base n digit.

2. Let w be a base n name and let d be a base n digit. Then $(wd)^s = w(d^+)$
   if d is not the biggest base n digit, and $(wd)^s = (w)^s 0$ if d is the biggest
   base n digit.

Example

1. In base 10, $(433)^s = 434$.
   Reason: since 3 is not the biggest base 10 digit, $(433)^s = 43(3^+) = 434$.

2. In base 10, $(499)^s = 500$.
   Reason: since 9 is the biggest base 10 digit we have $(499)^s = (49)^s 0 =$
   $(4)^s 00 = 4^+ 00 = 500$.

3. In base 5, $(44)^s = 100$.
   Reason: 4 is the biggest base 5 digit, so $(44)^s = (4)^s 0 = 100$.

Now we show our method of naming successors always works. That is, we
show that if z is any base n name, then z^s really does name the next number

after the one z names. Now, in base n, z names the number z_n, and so the *next* number is $(z_n)^+$. We must show it is this number that is named by z^s. But again in base n, z^s names the number $(z^s)_n$. So what we must show is that $(z^s)_n$ and $(z_n)^+$ are always the same.

Theorem 4.3.2 *In base n, let z be any name; then $(z^s)_n = (z_n)^+$.*

Proof By induction on the length of z.

If z is of length 1, it must be a digit, say d. Then we have two cases.

Case 1) d is not the biggest base n digit. Then d^+ is also a base n digit, so

$$
\begin{aligned}
(z^s)_n &= (d^s)_n & \text{since } z = d \\
&= (d^+)_n & \text{by definition} \\
&= d^+ & \text{since } d^+ \text{ is a digit} \\
&= (d_n)^+ & \text{since } d \text{ is a digit} \\
&= (z_n)^+ & \text{since } z = d.
\end{aligned}
$$

Case 2) d is the biggest base n digit. Then $d^+ = n$ (why?) so by Exercise 4.2.3, $d^+ = (10)_n$. Then

$$
\begin{aligned}
(z^s)_n &= (d^s)_n & \text{since } z = d \\
&= (10)_n & \text{by definition} \\
&= d^+ \\
&= (d_n)^+ & \text{since } d \text{ is a digit} \\
&= (z_n)^+ & \text{since } z = d
\end{aligned}
$$

We leave the induction step to you as an exercise. □

There is an important 'completeness' result that follows easily now.

Corollary 4.3.3 *Every whole number has a base n name.*

Proof 0 has a base n name since in every base 0 is a digit, and $(0)_n = 0$. The same is the case with 1.

Now suppose k has a base n name, we show the same is true of k^+. But, if z is a base n name for k, by the previous theorem z^s must be a base n name for k^+.

It follows, using induction, that every whole number has a base n name. □

Exercises

Exercise 4.3.1

1. In base 5, compute $(34)^s$.

2. In base 2, compute $(1011)^s$.

3. In base 12, compute $(te)^s$.

Exercise 4.3.2 Complete the proof of Theorem 4.3.2 by giving the induction step.

Exercise 4.3.3 Write out a list of names for the first 20 counting numbers in:

1. base 10;

2. base 5;

3. base 2;

4. base 12.

4.4   Comparing numbers

> Our intellects are a good deal sharpened up here, in some ways, and
> that is one of them. Numbers and signs and distances are so great, here,
> that we have to be made so we can *feel* them – our old ways of counting
> and measuring and ciphering wouldn't even give us an idea of them, but
> would only confuse us and oppress us and make our heads ache.
>
> > *–Captain Stormfield's Visit to Heaven*
> > Mark Twain, 1909

In Chapter Three we defined what it means for one whole number to be greater than another. In fact, place-value notation makes possible an easy *test* for the relative order of two numbers, by comparing their proper names. We just use the following informal rules, in which we assume everything is relative to base n, for some fixed n.

Rule 1 If two proper names are of different lengths, the longer name names the larger number.

Rule 2 If two proper names are the same length, then locate the first term (from the left) where the two names have different digits. The name with the bigger digit there names the bigger number.

Example This sort of thing should be quite familiar. In base 10, 231 names a bigger number than 99 since 231 is of length 3 while 99 is of length 2, which is smaller than 3. Likewise 7732 names a bigger number than 7719 because both are the same length, and agree in their first and second digits, but differ in their third, where 7732 contains the larger digit of the two.

If we make use of names that are not proper, then both rules are special cases of a single general rule.

General Rule If two names, proper or not, are the same length, then locate the first term from the left where the two names have different digits. The name with the bigger digit there names the bigger number.

It is obvious that Rule 2 follows from the General Rule, indeed it is simply the General Rule when we only consider proper names. But Rule 1 also follows because of Theorem 4.2.6 that says we can always adjoin 0's on the left, that is, if z is a name, then $0z$ and z name the same number. Now suppose we have two proper names, say z_1 and z_2, that are not the same length. Say z_2 is shorter. Rule 1 says z_1 must name the bigger number. But we can also argue as follows. Since z_2 is shorter than z_1, we can fill it out with 0's on the left to produce an equivalent name, $00\ldots0z_2$, that is the same length as z_1, though it is no longer a proper name. Now, z_1 was longer than z_2, so z_1 must consist of more than a single digit. And z_1 was a proper name, hence z_1 does not begin with 0. The name $00\ldots0z_2$ is the same length as z_1, and does begin with 0. Thus the two names differ in the very first term, and since 0 is the smallest digit (in any base) the General Rule says z_1 must name a bigger number than $00\ldots0z_2$, hence than z_2 itself.

Thus the correctness of the General Rule is all we need to establish. And we do this in the following theorem.

Theorem 4.4.1 *Suppose y and z are two base n names of the same length, and y begins with a bigger digit than z. Also, let w be some base n name. Then 1) $(y)_n > (z)_n$ and 2) $(wy)_n > (wz)_n$.*

Note Part 1 takes care of the case of two names differing at the very start. Part 2 takes care of the case where the two names, here wy and wz, agree for some initial string of digits, here string w, and then differ afterwards. The two cases could be lumped together if we wanted to allow w to be any string of base n digits including possibly the empty string.

Proof We use induction on the length of string y, or equivalently, z.

To start, suppose the length of y is 1. Then y and z must consist of single digits, say y consists of d_1, and z of d_2. Since the first term of y is a bigger digit than that of z, it must be that $d_1 > d_2$. Then we argue for Case 2 as follows (case 1 is much easier; check it).

$$
\begin{aligned}
(wy)_n &= (wd_1)_n \\
&= (w)_n \cdot n + d_1 \\
&> (w)_n \cdot n + d_2 \quad \text{(Theorem 3.5.2)} \\
&= (wd_2)_n \\
&= (wz)_n.
\end{aligned}
$$

Next, the induction step. Suppose the result is known for all names y and z of length k. And suppose now that we have particular names y and z of length k^+, with the first term of y being bigger than that of z. Since y is of length k^+, y must be of the form $y_0 a$ where y_0 is a name of length k, and a is a single digit. Thus $y = y_0 a$. Likewise since z is of length k^+, z must be of the form $z_0 b$ where z_0 is of length k and b is a digit. Thus $z = z_0 b$. Further, the first terms of y and y_0 must be the same, and the first terms of z and z_0 must be the same. Hence the first term of y_0 must be bigger than the first term of z_0. Now, again presenting only the Case 2 argument, we have

$$
\begin{aligned}
(wy)_n &= (wy_0 a)_n \\
&= (wy_0)_n \cdot n + a
\end{aligned}
$$

but a is a digit, hence $a \geq 0$, so using Theorem 3.5.2,

$$
\begin{aligned}
(wy)_n &\geq (wy_0)_n \cdot n + 0 \\
&= (wy_0)_n \cdot n
\end{aligned}
$$

Now y_0 and z_0 are names of length k and y_0 begins with a bigger digit than z_0, hence we can use the induction hypothesis. Then

$$
(wy_0)_n > (wz_0)_n.
$$

so by Exercise 3.4.1

$$
(wy_0)_n \geq (wz_0)_n^+.
$$

Then, using Theorem 3.5.3,

$$
(wy_0)_n \cdot n \geq (wz_0)_n^+ \cdot n
$$

and thus

$$
\begin{aligned}
(wy)_n &\geq (wz_0)_n^+ \cdot n \\
&= [(wz_0)_n + 1] \cdot n \\
&= (wz_0)_n \cdot n + 1 \cdot n \\
&= (wz_0)_n \cdot n + n.
\end{aligned}
$$

Now b is a base n digit hence by definition $n > b$. So by Theorem 3.5.2,

$$
\begin{aligned}
(wy)_n &> (wz_0)_n \cdot n + b \\
&= (wz_0 b)_n \\
&= (wz)_n.
\end{aligned}
$$

This concludes the induction step and completes the proof. $\square$

There is an important and easy consequence of all this. Recall that in Corollary 4.3.3 we showed every whole number had a base n name. If we allow improper as well as proper names, every whole number has many names. What we can show now, however, is that if we limit ourselves to proper names only, then numbers have *unique* names, relative to a fixed base, of course.

Corollary 4.4.2 *In base n notation every whole number has exactly one proper name.*

Proof Suppose some whole number had two proper base n names. By Rule 1 the two names must be of the same length. Then by Rule 2 they must be term by term the same, or else one would name a bigger number than the other. It follows that the two names must be identical. $\square$

Exercises

Exercise 4.4.1 Rank in order the following list of base 2 names: 11, 101, 111, 10, 1111, 1101, 1011.

4.5 Addition

> Clown: Let me see – every 'leven wether tods; every tod yields pound and odd shilling; fifteen hundred shorn, what comes the wool to? ...I cannot do't without counters.

> *–The Winter's Tale*
> Act IV Scene 17
> Shakespeare, 1610 or 1611 or so

There are convenient algorithms, using place-value notation, for doing all the elementary operations of arithmetic. In this section we discuss the usual one for adding. The problem is, given base n names for two whole numbers, to produce a base n name for their sum.

The usual addition algorithm tells us what to do with many-place names, provided we know what to do with one-place names, that is, with the digits themselves. So our first job is to produce an addition table for base n digits. The following arrangement, given for base 2, is useful.

+	0	1
0	0	1
1	1	10

The intention is that the bold face entry in the table above should represent $0 + 1$; more generally, the entry in row x and column y should name $x + y$.

The actual construction of such a table is quite simple. The definition given in Chapter Three, namely $x + 0 = 0 + x = x$ tells us how to fill in row 0 and column 0. Likewise, part of the definition in Chapter Two, namely $x + 1 = x^+$, together with the algorithm for naming successors, allows us to fill in the column headed by 1. Finally, the rest of the addition definition of Chapter Two, namely $x + y^+ = (x + y)^+$ tells us how to complete a column provided the one just to its left has been filled in. (This, of course, is not needed for base 2).

If you do Exercise 4.5.1 you may notice that these tables all share a feature with the one for base 2 presented earlier, that if a two-place name appears at all, the left-hand digit was a 1. This always happens.

Lemma 4.5.1 *In base n, when two digits are added, the result has a proper name which either is a single digit or consists of two digits the first of which is 1.*

Proof Let b be the biggest base n digit. Then, using the discussion of order in the previous section, it is enough to show that the sum of two base n digits is always $< (1b)_n$. Well, let d_1 and d_2 be two base n digits. Then $d_1 < n$ and $d_2 \leq b$, so

$$
\begin{aligned}
d_1 - d_2 \quad &< \quad n + b \\
&= \quad 1 \cdot n + b \\
&= \quad (1)_n \cdot n + b \\
&= \quad (1b)_n.
\end{aligned}
$$

$\square$

Definition 4.5.2 We say the sum of two base n digits *generates a carry* if a proper base n name for the sum consists of two digits (the first of which must be 1, by the Lemma).

Example In base 2, $1 + 1$ generates a carry. No other combination of base 2 digits does so.

Now we describe the algorithm for combining base n names consisting of more than single digits. The chief features of this familiar algorithm are these. We work from right to left. At each stage, we only consider one 'column' at a time. In that column we add the digits using our addition table, then we adjust the result, or not, depending on whether there was a carry generated by what we did in the previous column. If the result is (named by) a single digit we write it down, move left one column and repeat the process. If the result is (named by) more than one digit, we write down the 'units' digit, remember to carry, move left one place, and repeat.

Now, whenever we are finished with a column, we never work with it again. So when we move left one place it is as if we were 'throwing away' the old right-hand digits and shifting our attention to new right-hand digits of shorter names. Thus, in our formal statement of the algorithm below, it is enough to say what to do with right-hand digits only. Also, taking a carry into account is easy. The Lemma above essentially says that in adding two numbers we never have more than a 1 to carry. And adding in an extra 1 is simply taking successor.

Now, our *formal* statement of the addition algorithm.

Addition Algorithm Let $w_1 d_1$, $w_2 d_2$ and $w_3 d_3$ be base n names in which d_1, d_2, and d_3 are single digits. Then $(w_1 d_1)_n + (w_2 d_2)_n = (w_3 d_3)_n$ provided:

1. if $d_1 + d_2$ generates no carry, $d_3 = d_1 + d_2$, and $(w_3)_n = (w_1)_n + (w_2)_n$;

2. if $d_1 + d_2$ generates a carry, $(w_3)_n = [(w_1)_n + (w_2)_n]^+$.

Example We use this formal statement to compute (or name in base 2) $(101)_2 + (111)_2$. Since all our work is in base 2 notation, *we leave off all subscripts indicating base*, to make reading easier.

To begin, set

$$\underbrace{1 \quad 0 \quad 1}_{} \quad \underbrace{1 \quad 1 \quad 1}_{}.$$
$$\hspace{0.5em}\underset{w_1}{} \;\; \underset{d_1}{} \hspace{2em} \underset{w_2}{} \;\; \underset{d_2}{}$$

Now $d_1 + d_2 = 1 + 1 = 10$, using the base 2 addition table given earlier. Thus a carry is generated so case 2 applies. The result is $w_3 d_3$ where $d_3 = 0$ (the last term of 10) and $w_3 = (10 + 11)^+$. Thus the original problem of adding 3-place numbers (that is, numbers with 3-place names) is reduced to the problem of adding 2-place numbers.

We continue the calculation in a moment, but we pause to note that the customary arrangement of the work thus far, on a printed page, is

$$
\begin{array}{ccc}
 & 1 & \\
1 & 0 & 1 \\
1 & 1 & 1 \\
\hline
 & & 0
\end{array}
$$

Now we repeat the process to compute $10 + 11$, after which we can take the successor operation, or carry, into account to get $(10 + 11)^+$, which is what we really need. This time set

$$\underbrace{1 \quad 0}_{} \quad \underbrace{1 \quad 1}_{}$$
$$\underset{w_1}{} \;\; \underset{d_1}{} \hspace{2em} \underset{w_2}{} \;\; \underset{d_2}{}$$

(Note that we are re-using our variables with different values.) Now $d_1 + d_2 = 0 + 1 = 1$, again using the base 2 addition table. No carry is generated so we use Case 1. Then $10 + 11 = w_3 d_3$ where now $d_3 = 1$ and $w_3 = w_1 + w_2 = 1 + 1$. We still need to compute w_3, but we can state that $10 + 11 = w_3 1$. Then, using the successor algorithm of §3, (and uniformly writing a superscript $+$ for a superscript of s, as Theorem 4.3.2 allows) $(10 + 11)^+ = (w_3 1)^+ = w_3^+ 0$.

Earlier we determined that $101 + 111 = \ldots 0$ where $\ldots$ is a string of digits representing $(10 + 11)^+$. Well now we can fill this in a little more and say $101 + 111 = w_3^+ 00$ where $w_3 = 1 + 1$.

Again we pause to present the customary typographical arrangement of the work thus far. It is

$$
\begin{array}{ccc}
1 & 1 & \\
1 & 0 & 1 \\
1 & 1 & 1 \\
\hline
 & 0 & 0
\end{array}
$$

Finally, to complete the work we need w_3^+ where $w_3 = 1 + 1$. This time we go directly to the addition table. $1 + 1 = 10$, and by the successor algorithm, $(10)^+ = 11$. Thus $101 + 111 = w_3^+ 00 = 1100$. And the usual arrangement of the completed problem is:

$$
\begin{array}{cccc}
  & 1 & 1 &   \\
1 & 0 & 1 &   \\
1 & 1 & 1 &   \\
\hline
1 & 1 & 0 & 0
\end{array}
$$

Now we prove the correctness of the formal statement of the Addition Algorithm. In practice, it is the less formal typographical arrangement of the work that we assume you will use. We leave it to you to convince yourself that the usual 'by columns' arrangement is simply the Formal Algorithm in a handy abbreviated form.

Theorem 4.5.3 *The algorithm for base n addition given earlier is correct.*

Proof Curiously enough, we need some results from our discussion of division with remainder in Chapter Three. Specifically we need Theorem 3.10.1.

Suppose we know that $(w_1 d_1)_n + (w_2 d_2)_n = (w_3 d_3)_n$. What we must show is that the base n name $w_3 d_3$ meets the conditions set out in the statement of the Addition Algorithm given earlier. Now certainly

$$
\begin{aligned}
(w_3 d_3)_n &= (w_1 d_1)_n + (w_2 d_2)_n \\
&= [(w_1)_n \cdot n + d_1] + [(w_2)_n \cdot n + d_2] \\
&= [(w_1)_n \cdot n + (w_2)_n \cdot n] + [d_1 + d_2] \\
&= [(w_1)_n + (w_2)_n] \cdot n + [d_1 + d_2] \qquad (1)
\end{aligned}
$$

And also

$$
(w_3 d_3)_n = (w_3)_n \cdot n + d_3 \qquad (2)
$$

Now we have two cases.

Case 1) $d_1 + d_2$ generates no carry. Then $d_1 + d_2$ is a digit, so $d_1 + d_2 < n$. Now we use Theorem 3.10.1, taking x to be $(w_3 d_3)_n$ and y to be n. It tells us there are *unique* whole numbers q and r such that $(w_3 d_3)_n = q \cdot n + r$ and $r < n$. Now equation (1) says we can take $q = (w_1)_n + (w_2)_n$ and $r = d_1 + d_2$. Likewise, equation (2) says we can take $q = (w_3)_n$ and $r = d_3$. Consequently, it must be that $d_3 = d_1 + d_2$ and $(w_3)_n = (w_1)_n + (w_2)_n$. Thus the correctness of Case 1 of the Addition Algorithm is verified.

Case 2) $d_1 + d_2$ generates a carry. By Lemma 4.5.1 then, $d_1 + d_2 = (1d)_n$ for some base n digit d. Then, continuing equation (1) above,

$$
\begin{aligned}
(w_3 d_3)_n &= (w_1 d_1)_n + (w_2 d_2)_n \\
&= [(w_1)_n + (w_2)_n] \cdot n + [d_1 + d_2] \\
&= [(w_1)_n + (w_2)_n] \cdot n + (1d)_n \\
&= [(w_1)_n + (w_2)_n] \cdot n + (1 \cdot n + d) \\
&= [(w_1)_n + (w_2)_n + 1] \cdot n + d \\
&= [(w_1)_n + (w_2)_n]^+ \cdot n + d \qquad (3)
\end{aligned}
$$

Now the correctness of Case 2 of the Addition Algorithm follows, just as in Case 1), using Theorem 3.10.1 on equations (2) and (3). We leave any necessary details to the reader. □

Exercises

Exercise 4.5.1 Write out addition tables for the digits of:

1. base 5;

2. base 12;

3. base 10.

Exercise 4.5.2 Add, in base 2:

1. 101 + 110;

2. 111 + 11;

3. 1011 + 101.

Exercise 4.5.3 Add, in base 5:

1. 11 + 34;

2. 11 + 44;

3. 201 + 113;

4. 142 + 113.

Exercise 4.5.4 Add, in base 12:

1. 99 + 11;

2. te + 28;

3. tt0 + ee.

4.6   Subtraction

There are several methods of subtraction using place-value names. We take a brief look but leave much of the work to you.

Possibly the most common method used today is the one in which we work column by column, rearranging the 'top number' by borrowing when necessary. It is illustrated by the following base 10 example, in which we use the standard typographical arrangements. Suppose we have the problem:

$$
\begin{array}{r}
4 \ \ 3 \ \ 1 \\
- \ \ 1 \ \ 2 \ \ 9 \\
\hline
\end{array}
$$

We cannot subtract in the righthand column as it stands, so we 'borrow' to produce:

$$
\begin{array}{ccc}
  & 2 & 11 \\
4 & \cancel{3} & \cancel{1} \\
- \quad 1 & 2 & 9 \\
\hline
\end{array}
$$

Now we can calculate each column independently of the others. For example, $11 - 9 = 2$ because, by the base 10 addition table, $2 + 9 = 11$, and we recall how subtraction was defined in Chapter Three. Using the addition table backwards this way we get:

$$
\begin{array}{ccc}
  & 2 & 11 \\
4 & \cancel{3} & \cancel{1} \\
- \quad 1 & 2 & 9 \\
\hline
3 & 0 & 2 \\
\end{array}
$$

Now, a proper statement of the algorithm involved is simple, following the pattern begun for addition in the previous section.

Subtraction Algorithm Let $w_1 d_1$, $w_2 d_2$ and $w_3 d_3$ be base n names in which d_1, d_2 and d_3 are single digits. Suppose $(w_1 d_1)_n \geq (w_2 d_2)_n$. Then $(w_1 d_1)_n - (w_2 d_2)_n = (w_3 d_3)_n$ provided:

1. if $d_1 \geq d_2$ then $d_3 = d_1 - d_2$ and $(w_3)_n = (w_1)_n - (w_2)_n$;

2. if $d_1 < d_2$ then $d_3 = (n + d_1) - d_2$ and $(w_3)_n = [(w_1)_n - 1] - (w_2)_n$.

A second subtraction method sometimes taught is the borrowi-and-pay-back method. For example, consider again the base 10 problem:

$$
\begin{array}{ccc}
4 & 3 & 1 \\
- \quad 1 & 2 & 9 \\
\hline
\end{array}
$$

In this method, just like in the first one, we must borrow to modify the 1 in the righthand column. But now, the quantity borrowed to turn the 1 into 11 is 'paid back' to the *bottom* number. Thus the problem is rearranged into:

$$
\begin{array}{ccc}
  & & 11 \\
4 & 3 & \cancel{1} \\
- \quad 1 & \cancel{2}^3 & 9 \\
\hline
\end{array}
$$

Then we work column by column as before.

Finally there is one other method of subtraction that is rarely taught in elementary schools today, but which has considerable use in computer design. For base 10 it involves *9's complements* and goes as follows.

In base 10, the *9's complement* of a digit d is that digit one can add to d to produce 9. For example, the 9's complement of 0 is 9, of 1 is 8, and so on. Note that every base 10 digit has exactly one 9's complement.

Now, consider again the base 10 problem

$$
\begin{array}{r}
4\ \ 3\ \ 1 \\
-\ \ 1\ \ 2\ \ 9 \\
\hline
\end{array}
$$

In this method we begin by replacing each digit in the bottom number by its 9's complement. Then we *add*. We get

$$
\begin{array}{r}
4\ \ 3\ \ 1 \\
+\ \ 8\ \ 7\ \ 0 \\
\hline
1\ \ 3\ \ 0\ \ 1
\end{array}
$$

Next we take the *successor* of the result. In this case we get 1302. No matter what the base 10 subtraction problem we started with, the result of this step will begin with the digit 1. Now delete it. We get 302 and this is the answer to the original *subtraction* problem.

One more thing needs to be said, however. This method only works if the two names are of the same length. If the bottom one is shorter, adjoin 0's on the left using Theorem 4.2.6 until both names are the same length. Thus to compute:

$$
\begin{array}{r}
21648 \\
-\ \ \ \ \ 929 \\
\hline
\end{array}
$$

we would use the algorithm on

$$
\begin{array}{r}
21648 \\
-\ \ 00929 \\
\hline
\end{array}
$$

Note that this subtraction method actually converts the problem into one of addition. Thus no separate notion of borrowing is necessary.

Rather than giving a formal statement of this algorithm and proving it correct, we confine ourselves to a special case. In what follows, we work in *base 10* and we work with names of *length 3*. This is to keep the notational clutter down. We think all the essential ideas of the method are adequately displayed.

Suppose abc and efg are two base 10 names of length 3 (so a, b, c, e, f and g are base 10 digits), and suppose $(abc)_{10} \geq (efg)_{10}$. We use the 9's complement method on the subtraction problem:

$$
\begin{array}{r}
a\ \ b\ \ c \\
-\ \ e\ \ f\ \ g \\
\hline
\end{array}
$$

Let us write $\bar{d}$ for the 9's complement of the digit d. Thus $\bar{2} = 7$, for example. Obviously $d + \bar{d} = 9$.

The first step in the subtraction algorithm is to replace each digit of the bottom number by its 9's complement, and add.

$$
\begin{array}{ccc}
a & b & c \\
\hline
+ \quad e & f & g \\
\hline
\end{array}
$$

We get $(abc)_{10} + (\overline{e}\,\overline{f}\,\overline{g})_{10}$. Next we take successor, getting

$$(abc)_{10} + (\overline{efg})_{10} + 1 \qquad\qquad (*)$$

Now, by our methods of *addition*,

$$
\begin{array}{ccc}
\overline{e} & \overline{f} & \overline{g} \\
+ \quad e & f & g \\
\hline
9 & 9 & 9 \\
\end{array}
$$

hence $(\overline{e}\,\overline{f}\,\overline{g})_{10} = (999)_{10} - (efg)_{10}$ by definition of subtraction. Thus the item labeled $(*)$ above is equal to $(abc)_{10} + [(999)_{10} - (efg)_{10}] + 1$. Using various properties of subtraction from Chapter Three, this can be rearranged into $[(abc)_{10} - (efg)_{10}] + [(999)_{10} + 1]$. Now, our methods of addition tell us that $(999)_{10} + 1 = (1000)_{10}$ hence, so far we have:

$$[(abc)_{10} - (efg)_{10}] + (1000)_{10}. \qquad\qquad (**)$$

The final step of the algorithm is to delete the initial digit 1. Equivalently (for our special case of starting with 3 place numbers), we reduce things by $(1000)_{10}$. This turns item $(**)$ into $(abc)_{10} - (efg)_{10}$. In other words, the 9's complement algorithm produced the subtraction result we were after.

The method works with any base, of course. In base 5 we would use 4's complements; in base 12, 11's complements. The method is at its best in base 2. There one would use 1's complements. But the 1's complement of 0 is 1, and of 1 is 0. Thus the step of introducing 1's complements into the bottom number simply amounts to switching 0's and 1's. In fact, this is how computers (which work in base 2) generally subtract.

Exercises

Exercise 4.6.1 Carry out the base 10 problem $(431)_{10} - (129)_{10}$, but using the formal statement of the subtraction algorithm.

Exercise 4.6.2 In case 2) of the Subtraction Algorithm two subtractions occur.

1. Prove that if $d_1 < d_2$ then $(n + d_1) - d_2$ is a digit.

2. Prove that if $d_1 < d_2$ then $[(w_1)_n - 1] - (w_2)_n$ is defined.

Hint: For each part, assume the contrary and derive a contradiction.

Exercise 4.6.3 Prove the Subtraction Algorithm is correct.

Exercise 4.6.4 Use the informal version, by columns, of the Subtraction Algorithm to compute the following.

1. In base 5:

   a) $14 - 12$,

   b) $23 - 4$,

   c) $211 - 102$,

   d) $211 - 22$;

2. In base 12 $4t9e0 - t371$.

3. In base 2 $10011101 - 1100111$.

Exercise 4.6.5 Give a formal statement of the subtraction algorithm involving borrowing and paying back.

Exercise 4.6.6 Prove the correctness of the Algorithm given in answer to Exercise 4.6.5.

Exercise 4.6.7 Use the borrow and pay back method of subtraction to re-do Exercise 4.6.4.

Exercise 4.6.8 In base n, for any digit d, show there is exactly one digit $\bar{d}$ such that $d + \bar{d} = n - 1$.

Exercise 4.6.9 Use the 2's complement method to compute, in base 2:

1. $10011101 - 1100111$;

2. $10000 - 111$

3. $100001 - 11100$.

4.7   Multiplication

> Bunthorpe: Do you know what it is to yearn for the Indefinable, and yet to be brought face to face, daily, with the Multiplication Table?
>
> –*Patience*,
> W.S.Gilbert, 1881

There is a simple, familiar algorithm for multiplying numbers by using their place-value names. And just as for addition, what it does is reduce a problem involving many-digit names to a collection of problems each involving single digits. How to multiply digits is simply learned, once and for all. So we must begin by creating a base n *multiplication table* for each n we are interested in. As one might expect, the simplest is for base 2, and we present it by way of example.

$\times$	0	1
0	0	0
1	0	1

To create a table for any base, all that is needed beyond addition, is the definition of multiplication itself. That is, we need:

$$\begin{aligned} x \cdot 0 &= 0 \cdot x = 0 \quad \text{(Chapter Three)} \\ x \cdot 1 &= x \\ x \cdot y^+ &= x \cdot y + x \quad \text{(Chapter Two)} \end{aligned}$$

Now, the workings of the algorithm itself really all come down to a single fact, and that in turn rests on little more than the distributive law. The fact we need is this.

Theorem 4.7.1 *Let wd be a base n name, in which d is a single digit. And let x be any whole number. Then $(wd)_n \cdot x = (w)_n \cdot x \cdot (10)_n + d \cdot x$.*

Proof By definition $(wd)_n = (w)_n \cdot n + d$. Also Exercise 4.2.3 says $n = (10)_n$. The Theorem now follows by using the distributive law. $\square$

In words, this essentially says that when we multiply by x we can do so digit by digit, but each shift to the left one place sends things up by a factor of $(10)_n$. And we remind you that Exercise 4.2.4 says that multiplying by $(10)_n$ is just adjoining a 0 to the right hand end of the name. Now let us see how this works in practice.

Suppose we have a base 5 multiplication problem in which one of the numbers has a single digit name. We will take care of the more general case shortly. And we assume you have a base 5 multiplication table handy. Say we have the problem $(243)_5 \cdot 4$. By repeated use of the theorem above, and the multiplication table, we get:

$$\begin{aligned} (243)_5 \cdot 4 &= (24)_5 \cdot 4 \cdot (10)_5 + 3 \cdot 4 \\ &= (24)_5 \cdot 4 \cdot (10)_5 + (22)_5 \\ &= [(2)_5 \cdot (10)_5 + 4 \cdot 4] \cdot (10)_5 + (22)_5 \\ &= (2)_5 \cdot 4 \cdot (10)_5 \cdot (10)_5 + 4 \cdot 4 \cdot (10)_5 + (22)_5 \\ &= (13)_5 \cdot (10)_5 \cdot (10)_5 + (31)_5 \cdot (10)_5 + (22)_5 \\ &= (1300)_5 + (310)_5 + (22)_5 \end{aligned}$$

Now we have reduced things to a base 5 addition problem. Here we leave out all details since they were covered in §5. The sum is $(2132)_5$.

This work can be arranged on a printed page in more familiar fashion thus:

$$
\begin{array}{rccc}
       & 2 & 4 & 3 \\
\times &   &   & 4 \\
\hline
       &   & 2 & 2 \\
       & 3 & 1 & 0 \\
     1 & 3 & 0 & 0 \\
\hline
     2 & 1 & 3 & 2 \\
\end{array}
$$

In base 10, though, the *partial products* are seldom written down, and the addition is done in the head.

Now, suppose we have two many-digit names to work with. Say we have the base 5 problem $(243)_5 \cdot (42)_5$. Well, we again use the theorem. But, relying on the commutativity of multiplication, we use it this time to 'break up' the right-hand factor. We get $(243)_5 \cdot (42)_5 = (243)_5 \cdot 4 \cdot (10)_5 + (243)_5 \cdot 2$. Now we just worked out that $(243)_5 \cdot 4 = (2132)_5$. In a similar way we can calculate that $(243)_5 \cdot (2) = (1041)_5$. Hence we have $(243)_5 \cdot (42)_5 = (2132)_5 \cdot (10)_5 + (1041)_5 = (21320)_5 + (1041)_5$. And this sum works out to $(22411)_5$. Of course, the customary arrangement of this is:

$$
\begin{array}{rccccc}
       &   & 2 & 4 & 3 \\
\times &   &   & 4 & 2 \\
\hline
       & 1 & 0 & 4 & 1 \\
     2 & 1 & 3 & 2 & 0 \\
\hline
     2 & 2 & 4 & 1 & 1 \\
\end{array}
$$

(though in the second partial product, the final 0 is not generally written, but rather one just shifts everything left one place.)

Exercises

Exercise 4.7.1 Write out multiplication tables for the digits of:

1. base 5;

2. base 12;

3. base 10.

Exercise 4.7.2 In base 5, calculate $24 \cdot 32$ and $103 \cdot 21$.

Exercise 4.7.3 In base 2, calculate $1101 \cdot 111$ and $10101 \cdot 11110$.

Exercise 4.7.4 In base 12, calculate:

1. $99 \cdot 99$;

2. *tet* $\cdot 28$;

3. *tt* $\cdot$ *ee*.

4.8 Division

The usual division algorithm using place-value names is quite easy to present on a theoretical level. It is nothing more than the multiplication algorithm run in reverse. Consequently all we do is give an example. That should suffice.

If we want to divide y into x, where $y \neq 0$, Theorem 3.10.1 says it can be done, and in only one way. That is, there is a unique quotient q and a unique remainder r meeting the conditions:

$$x = q \cdot y + r$$

$$r < y.$$

Now in the proof of that theorem it is clearly seen that the right choice for q is simply the *largest* number y such that $q \cdot y \leq x$. This is the guiding principle behind what follows. Suppose we want, in base 10, to divide 528 into 192200. What we will do is find the largest number q such that $q \cdot 528 \leq 192200$. All the work that follows is in base 10 notation.

Our rules for comparing numbers easily tell us that 1000 is the smallest 4-place number. (That is, it is the smallest number whose proper name has 4 places.) And $1000 \cdot 518 = 10 \cdot 10 \cdot 10 \cdot 528 = 528000$ (using Exercise 4.2.4). This is bigger than 192200. Consequently q can not be 4-place; it must be 3-place or less.

Now we can think of what follows as a puzzle. We want digits to put in the blanks of the following multiplication so that the answer is as large as possible while not being larger than 192200.

```
              5  2  8
×             -  -  -
        .  .  .  .  .
        .  .  .  .  .
        .  .  .  .  .
        .  .  .  .  .
```

Our rules for comparing numbers have us begin by looking at the left-hand digits first. Consequently, if we want the *largest* number possible, we should begin by filling the left-hand blank (called the hundreds place, in this case) by the largest digit we can.

We can proceed by trial and error. Say we try 4. Then, by the methods of multiplication discussed in the previous section, we can be certain of at least part of the multiplication arrangement, namely:

```
              5  2  8
×             4  -  -
        .  .  .  .  .
        .  .  .  .  .
     2  1  1  2  0  0
```

Now, whatever the other partial products are, they will all add up to at least 211200, which is too big, since we are aiming at 192200.

Continuing trial and error shows 3 is the largest digit we can use. So far our (incomplete) multiplication looks like this:

$$
\begin{array}{r}
5\ \ 2\ \ 8 \\
\times \qquad\quad 3\ \ \_\ \ \_ \\
\hline
\cdot\ \ \cdot\ \ \cdot\ \ \cdot\ \ \cdot\ \ \cdot \\
\cdot\ \ \cdot\ \ \cdot\ \ \cdot\ \ \cdot\ \ \cdot \\
\hline
1\ \ 5\ \ 8\ \ 4\ \ 0\ \ 0
\end{array}
$$

Now, this partial product accounts for 158400 of the 192200 we had to work with. Consequently we have

$$
\begin{array}{r}
1\ \ 9\ \ 2\ \ 2\ \ 0\ \ 0 \\
-\ \ 1\ \ 5\ \ 8\ \ 4\ \ 0\ \ 0 \\
\hline
3\ \ 3\ \ 8\ \ 0\ \ 0
\end{array}
$$

left to distribute between the other two partial products.

We now move right one place and try to fill the next digit. We do so by repeating the whole process all over again, but now our goal is to be as big as possible without exceeding 33800, since this is all we have left to take care of. Now, let us simply rearrange the work thus far into a more familiar pattern.

$$
\begin{array}{r}
3 \\
528\overline{)192200} \\
\underline{1584} \\
33800
\end{array}
$$

You should recognize all the parts here as having occurred earlier, though somewhat scattered about. And we think this is far enough to carry this example. Our point is: the usual division algorithm is nothing more than a convenient typographical arrangement of the usual multiplication algorithm 'run backwards'.

Division in base 2 is particularly simple as, at each stage, we have only 0 and 1 to try.

Exercises

Exercise 4.8.1 In base 2:

1. divide 101 into 111001;

2. divide 101 into 111111.

3. For parts 1 and 2, check your work by showing that multiplying by the quotient and adding the remainder produces what it should.

4.9 Changing bases

We have been considering many bases, so a simple method of converting from one base to another is of interest. If the numbers involved are not very big, one can convert by simply counting in two bases simultaneously, thus generating a conversion chart or table for the bases in question. It was by this method that we concluded $(11)_2 = (3)_{10}$ in Exercise 4.2.5. Similarly, if you did Exercise 4.3.3, you know how to convert between bases 10, 5, 2 and 12 for the first 20 numbers. What we need now are methods that work conveniently when large numbers are involved.

Actually, we present two methods of converting a base n name into an equivalent base k name. In both, a certain amount of calculation is necessary. The difference between the two methods, on a practical level, is that in one it is most natural to conduct the calculations using base k notation, and in the other, using base n notation.

Converting from base n to base k Method One To use this method we must first know how to name each number $\leq n$ in base k. That is, we need base k names for each base n *digit* and for the base $n = (10)_n$ itself. We can do this by the counting method outlined above.

Thus going from base 12 to base 10 we need the following information.

$$
\begin{array}{llll}
(0)_{12} & = & (0)_{10} & (6)_{12} & = & (6)_{10} \\
(1)_{12} & = & (1)_{10} & (7)_{12} & = & (7)_{10} \\
(2)_{12} & = & (2)_{10} & (8)_{12} & = & (8)_{10} \\
(3)_{12} & = & (3)_{10} & (9)_{12} & = & (9)_{10} \\
(4)_{12} & = & (4)_{10} & (t)_{12} & = & (10)_{10} \\
(5)_{12} & = & (5)_{10} & (e)_{12} & = & (11)_{10} \\
\end{array}
$$

$$(10)_{12} = (12)_{10}$$

Now, continuing our presentation of Method One, we give the *reduction step*. This reduces the problem of converting a given base n name into base k to the problem of converting a *shorter* base n name into base k. Continued applications of this reduction step brings us down to the level of converting length 1 names. But these are just the base n digits, and we are assuming we know how to convert those.

Let wd be a base n name, where d is a base n digit. Essentially by definition we have:

$$(wd)_n = (w)_n \cdot (10)_n + (d)_n.$$

Now, suppose we had *base k* names for all the terms on the right in this equation, namely for $(w)_n$, for $(10)_n$ and for $(d)_n$. Then we could combine them as directed, using the adding and multiplying algorithms that work with base k names. The result would be a base k name for $(wd)_n$. But, in fact, we are assuming we have base k names for d (a base n digit) and for $(10)_n$. All we need is a base k name for $(w)_n$. But w is a *shorter* name than wd, the name we began with. This is the reduction step we spoke of earlier.

Example Convert $(2te)_{12}$ to base 10. We use the conversions above. Now:

$$
\begin{aligned}
(2te)_{12} &= (2t)_{12} \cdot (10)_{12} + (e)_{12} \\
&= (2t)_{12} \cdot (12)_{10} + (11)_{10}
\end{aligned}
$$

So the problem has been simplified to the conversion of $(2t)_{12}$. Iterating the process we have:

$$
\begin{aligned}
(2t)_{12} &= (2)_{12} \cdot (10)_{12} + (t)_{12} \\
&= (2)_{10} \cdot (12)_{10} + (10)_{10}
\end{aligned}
$$

and now, carrying on calculations in base 10 we get

$$
\begin{aligned}
&= (24)_{10} + (10)_{10} \\
&= (34)_{10}.
\end{aligned}
$$

Consequently

$$
\begin{aligned}
(2te)_{12} &= (2t)_{12} \cdot (12)_{10} + (11)_{10} \\
&= (34)_{10} \cdot (12)_{10} + (11)_{10} \\
&= (408)_{10} + (11)_{10} \\
&= (419)_{10}.
\end{aligned}
$$

Notice that all calculations were conducted in base 10 notation, that is, in the base we were *converting to*.

Converting from base n to base k Method Two
This method is a sort of dual to the previous one. This time we assume we have *base n* names for the base k digits and for k itself. Then, as before, there is a reduction step, showing how to turn a conversion problem into a simpler conversion problem; this time into one involving smaller numbers.

Suppose we have a base n name, say z. The problem is to find a base k name for the number $(z)_n$. Now, say wd is such a name, where d is a base k digit. That is, suppose $(z)_n = (wd)_k$. We show how to determine d.

By definition, $(wd)_k = (w)_k \cdot k + d$. Since d is a base k digit, $d < k$. But this is exactly the setup for long division. Theorem 3.10.1 tells us that $(w)_k$ must be the *quotient*, and d the *remainder* on dividing $(wd)_k$ by k. But quotient and remainder are independent of the system of notation in which calculations are carried out. Let us do the calculations in *base n*. That is, we divide $(wd)_k$, which is just $(z)_n$, by k, and we are assuming we have a base n name for that. The remainder must be less than k, hence must be a base k digit, though we get a base n *name* for it. But we are assuming knowledge of how the base k digits are named in base n. Thus, d has been determined.

What is left? We wanted the string of digits wd. We have determined d itself. But we also know the value $(w)_k$; it is simply the *quotient* of the division carried out above. Thus the problem has been simplified, since $(w)_k$ is a smaller number than $(wd)_k$ was.

In short, divide the number by k; the remainder gives us the right hand digit in base k. But the *quotient*, $(w)_k$ above, is the 'rest' of the number, so we may iterate the process with that in place of the original value. Thus we conduct successive divisions, each time dividing k into the quotient of the previous step. The successive remainders give us the desired digits of the base k name, in right-to-left order.

Example Convert $(419)_{10}$ to base 12. Well, we simply keep dividing successive quotients by 12, doing all the work in base 10 notation. We get:

$$
\begin{array}{r}
34 \\
12)\overline{419} \\
\underline{36} \\
59 \\
\underline{48} \\
11 \\
\end{array}
$$

$$
\begin{array}{r}
2 \\
12)\overline{34} \\
\underline{24} \\
10 \\
\end{array}
$$

$$
\begin{array}{r}
0 \\
12)\overline{2} \\
\underline{0} \\
2 \\
\end{array}
$$

Now the successive remainders are 11, 10, 2. Writing these as base 12 digits, we have e, t, 2. Arranging these from right to left we get the base 12 name $2te$. Thus $(419)_{10} = (2te)_{12}$.

Notice that all calculations were conducted in base 10 notation; that is, in the base we were *converting from*.

Exercises

Exercise 4.9.1 Find base 10 names for:

1. $(101101)_2$;

2. $(7te)_{12}$.

Exercise 4.9.2 Convert $(324)_{10}$ to:

1. base 5;

2. base 2.

Exercise 4.9.3 Use the method of repeated division and convert $(324)_{10}$ to base 5, and to base 2.

The examples presented above involved base 10, which is the most familiar to us. The conversion methods presented require calculating ability in either the base we are coming from or the base we are going to. If neither of them is base 10, you may find it simplest to go through 10 as an intermediate step. That is, to go from base n to base k, first go from n to 10 using Method One, then from 10 to k using Method Two. This way all calculations can be carried out in base 10 notation. There is no theoretical necessity for going through base 10; it is merely a matter of convenience.

Exercise 4.9.4

1. Convert $(241)_5$ into base 2.

2. Convert $(2et)_{12}$ into base 5.

Exercise 4.9.5 Carry out an addition, subtraction, multiplication or division problem in some base other than 10, then check your work by converting the problem and its answer into base 10 and seeing if it is correct there.

The next several exercises are almost recreational in character. Several are taken from Chrystal's classic work, *Algebra*, subtitled, *an elementary text-book for the higher classes of secondary schools and for colleges*, a book first published in 1886, and with additions, in 1904.

Exercise 4.9.6 Expressed in base n, $(79)_{10}$ becomes $(142)_n$. Find n.

Exercise 4.9.7 A 3 digit name in base 7 notation has its digits reversed when translated into base 9. Find the digits.

Exercise 4.9.8

1. Show, in base 10, the difference between the square of a two digit number, and the square of the number formed by reversing the digits, is divisible by 99.

2. Generalize this to base n.

Exercise 4.9.9

1. Show the following base 10 number trick is correct. Take any 3 digit number whose first digit is bigger than its last digit (say 532). Reverse the digits and subtract the result from the original number (in our example, $532 - 235 = 297$). Take the result, reverse the digits, and add (in our example, $297 + 792 = 1089$). Show the result is always 1089.

2. Show that in any base the result is always given by $1100 - 11$.

CHAPTER 5

The Rational Number System

Giulia: It's quite simple. Observe. Two husbands have managed to acquire three wives. Three wives — two husbands. (Reckoning up) That's two-thirds of a husband to each wife.

Tessa: O Mount Vesuvius, here we are in arithmetic! My good sir, one can't marry a vulgar fraction!

Giulia: You've no right to call me a vulgar fraction.

Marco Palmieri: We are getting rather mixed.

> –The Gondoliers
> W. S. Gilbert, 1889

5.1   Introduction

We have, so far, discussed numbers for *counting*. Now we discuss numbers for *measuring*. It turns out that this is a new kind of number, but one that can be defined in terms of whole numbers, so new axioms are not needed, only appropriate definitions. And to motivate these definitions we discuss measuring, just as in Chapter Two we discussed counting.

In the real world, exact measurement has no meaning. Dimensions change with temperature. Yardsticks are subject to wear. We do the best we can, but knock a few atoms off the edge of a chair and we will never notice it. All real measurements are only approximate things. So, if we want to study measurement from a mathematical point of view, we are going to have to idealize the situation. We want to talk about objects that have fixed, well-defined lengths. We want to talk about objects that can be divided as finely as we want, unlike real yardsticks which, on too fine a division, become sawdust. What we actually talk about is *line segments*.

We assume you have some intuitive feeling for the geometric behavior of line segments, and we rely on this in our discussions. But these discussions are only meant to motivate our definitions. They are not proofs, and so no formal development of geometry is given. The properties we claim for addition, say, are *true* because we prove them, based on our definition of addition. But addition is of *interest* to us because of its intuitive connection with measurement.

For this chapter, all whole number names are base 10, unless another base is specifically indicated.

5.2   Fractions

> ...though from the magnitude of the figure it might at first deceive a landsman, yet the slightest consideration will show that though seven hundred and seventy-seven is a pretty large number, yet when you come to make a teenth of it, you will then see, I say, that the seven hundred and seventy-seventh part of a farthing is a good deal less than seven hundred and seventy-seven gold doubloons; and so I thought at the time.
>
> –Moby Dick
> Herman Melville, 1851

We begin our discussion of measuring line segments by choosing a *standard length*. Say we choose:

We call this a *Unit length*, or just a *Unit*.

If we say line segment A is 3 Units, we simply mean 3 of these Unit lengths exactly fit into A.

A

Similarly for 4 Unit line segments, 5 Units, and so on.

But, for measuring certain line segments, our Unit may be too gross. Some line segments may fall between, say, 3 Units and 4 Units. We need a finer measure. So, we divide our Unit up into smaller pieces. We take them all to be of the same size, but the number of them can be as large as we want. Say we divide our Unit into 5 pieces.

Unit

Then each piece is a 5th (of the Unit) Similarly for 6th, 7th, etc. (English usage has us write 3rd instead of 3th, and half or 2nd for 2th.)

Now, in effect, we can use 5th as a new standard length. We may say line segment B is 7 5th's if exactly 7 of our 5th's fit into it.

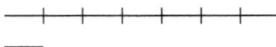

B
5th

Let us say a line segment, C, is *measurable* if C is k n^{th}'s (of our Unit) for some k and some n. We leave open the question of whether all line segments are measurable or not, and we concentrate on those which do turn out to be.

Suppose line segment D is measurable. Say it is 9 7th's. We need a convenient way of saying this. The two numbers 9 and 7 must be involved, and we must be able to tell which number represents divisions of our Unit and which is the number of those divisions which fit into D. The world at various times has used many different devices for indicating this in writing. One ancient Greek method did the equivalent of writing 9 7'. In English, we would write "nine sevenths." The technical symbol generally adopted in the world today for this purpose is the *fraction*, in this case $\frac{9}{7}$. (Sometimes this is written as 9/7, for typographical convenience.) Mathematically this is just an *ordered pair*, consisting of 9 and 7 in a particular order.

In the fraction $\frac{9}{7}$, it is customary to call 7 the *denominator*, indicating the *denomination*, or size, of our measure, in this case 7th. The 9 is called the *numerator*, indicating the *number* of 7th's which fit into D. The word *fraction* itself is akin to *fracture*. Indeed, at one time fractions were also called *broken numbers*, for obvious reasons, contrasting with *whole numbers*.

In summary, if we have the situation

we say D is $\frac{9}{7}$ Unit.

When we studied numbers for counting we found that 0 played a special role. It is reasonable to ask how it behaves in fractions.

Consider first the symbol $\frac{0}{7}$; what meaning can be assigned to it? It tells us to divide our Unit into 7 parts, but to use *none* of them. $\frac{0}{7}$ is the appropriate symbol to use if it turns out we have nothing to measure, just as we use 0 when we have nothing to count. By the way, $\frac{0}{8}$ or $\frac{0}{6}$ would do quite as well as $\frac{0}{7}$; they all tell us we don't need any of our lengths to measure what we have, that is, we have nothing to measure.

Now consider the symbol $\frac{c}{c}$. It tells us to divide our Unit length into 0 parts. But our Unit is in 1 piece to begin with; if we divide it up we will get more than 1 piece. No matter how we chop it up, we can never divide our Unit into 0 pieces. The symbol $\frac{9}{0}$ tells us to do something which is not possible, so we say it is meaningless, and we do not allow it as a fraction.

Definition 5.2.1 A *fraction* is an ordered pair of whole numbers, by custom written $\frac{a}{b}$, in which $b \neq 0$.

A convention that will save us much writing: whenever we write $\frac{a}{b}$ we mean it is a fraction, and so $b \neq 0$. We will not use the symbol otherwise.

Exercises

Exercise 5.2.1 Suppose line segment A is 3 Units long. How would this be said using fractions?

5.3   Equivalence of fractions

Suppose line segment A is $\frac{3}{4}$ Unit. That is,

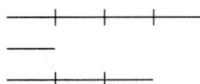

$$\begin{array}{ll}
\underline{\quad|\quad|\quad|\quad} & \text{Unit} \\
\underline{\qquad} & 4^{th} \\
\underline{\quad|\quad|\quad} & A
\end{array}$$

Now, suppose we divide our Unit into twice as many parts (by dividing each 4th into two). Then twice as many of them will be needed to fill out A.

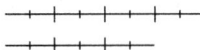

$$\begin{array}{ll}
\underline{\,|\,|\,|\,|\,|\,|\,|\,} & \text{Unit} \\
\underline{\,|\,|\,|\,|\,|\,} & A
\end{array}$$

This means that A is also $\frac{6}{8}$ Unit.

Similarly, if we divide our Unit into three times as many parts (by dividing each 4th into 3 parts), we will need three times as many of them to make up A.

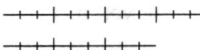

$$\begin{array}{ll}
\underline{\,|\,|\,|\,|\,|\,|\,|\,|\,|\,|\,|\,|\,} & \text{Unit} \\
\underline{\,|\,|\,|\,|\,|\,|\,|\,|\,|\,} & A
\end{array}$$

Then A is also $\frac{9}{12}$ Unit.

More generally, if we divide our Unit into n times as many parts (getting $4 \cdot n$ of them) we will need n times as many of them to make up A (in this case $3 \cdot n$ of them). Then A is also $\frac{3 \cdot n}{4 \cdot n}$ Unit. So, when we measure A, there is no single fraction we must come up with; $\frac{3}{4}$ or $\frac{6}{8}$ or $\frac{9}{12}$ or $\frac{3 \cdot n}{4 \cdot n}$ all will serve. More generally yet, if line segment B is $\frac{a}{b}$ Unit, it is also $\frac{n \cdot a}{n \cdot b}$ Unit, for any counting number n.

Let us, informally, say two fractions are *interchangeable* if they represent measurements of the same line segment. Thus $\frac{6}{8}$ and $\frac{9}{12}$ are interchangeable. And by arguments like the above, $\frac{a}{b}$ and $\frac{n \cdot a}{n \cdot b}$ are interchangeable. What we need is a general test for being interchangable, so we can recognize it by just looking at the fractions, without going back to line segments.

Suppose $\frac{a}{b}$ and $\frac{c}{d}$ are interchangeable fractions. Then they both turn up in measuring the same line segment, say B.

$$\begin{array}{ll} \underline{\hspace{3cm}} & \text{Unit} \\ \underline{\hspace{3cm}} & \text{B} \end{array}$$

Thus B is $\frac{a}{b}$ Unit, and also $\frac{c}{a}$ Unit. Now $\frac{a}{b}$ and $\frac{n \cdot a}{n \cdot b}$ are interchangeable for any counting number n, in particular, for $n = d$. That is, $\frac{a}{b}$ and $\frac{d \cdot a}{d \cdot b}$ are interchangeable, so B is also $\frac{a \cdot d}{b \cdot d}$ Unit. Similarly $\frac{c}{d}$ and $\frac{b \cdot c}{b \cdot d}$ are interchangeable, so B is also $\frac{b \cdot c}{b \cdot d}$ Unit. B is both $\frac{a \cdot d}{b \cdot d}$ Unit and $\frac{b \cdot c}{b \cdot d}$ Unit. Then if we divide our Unit into $b \cdot d$ pieces, *either* $a \cdot d$ or $b \cdot c$ of them will exactly fit into B. But intuition tells us only one quantity of them will make up B. So, it must be that $a \cdot d = b \cdot c$.

We have given an intuitive argument that, if $\frac{a}{b}$ and $\frac{c}{d}$ are interchangeable, then $a \cdot d = b \cdot c$. Now let us argue that this works backwards as well.

Suppose $a \cdot d = b \cdot c$, and consider the fractions $\frac{a}{b}$ and $\frac{c}{d}$. Now, $\frac{a}{b}$ and $\frac{a \cdot d}{b \cdot d}$ are interchangeable. Similarly, $\frac{c}{d}$ and $\frac{b \cdot c}{b \cdot d}$ are interchangeable. But, since $a \cdot d = b \cdot c$, then $\frac{a \cdot d}{b \cdot d}$ and $\frac{b \cdot c}{b \cdot d}$ are the same fractions. Then $\frac{a}{b}$ and $\frac{c}{d}$ are interchangeable with the same thing, so, intuitively, they must be interchangeable with each other.

We have *informally* shown that $\frac{a}{b}$ and $\frac{c}{d}$ will be interchangeable precisely if $a \cdot d = b \cdot c$. This provides a reasonable definition of a *formal* counterpart to interchangeability.

Definition 5.3.1 Let $\frac{a}{b}$ and $\frac{c}{d}$ be fractions. We say $\frac{a}{b}$ and $\frac{c}{d}$ are *equivalent* if $a \cdot d = b \cdot c$. If $\frac{a}{b}$ and $\frac{c}{d}$ are equivalent we write $\frac{a}{b} \sim \frac{c}{d}$.

Remark We only write $\frac{a}{b} = \frac{c}{d}$ if $\frac{a}{b}$ and $\frac{c}{d}$ are the *same* fraction, that is, if $a = c$ and $b = d$. Then $\frac{1}{2} = \frac{2}{4}$ is *not* true. But $\frac{1}{2} \sim \frac{2}{4}$ is true since $1 \cdot 4 = 2 \cdot 2$. Equivalence is of interest to us because it intuitively corresponds to the notion of interchangeability, which has to do with measurement. But the definition of equivalence is in terms of whole numbers and multiplication, and it is what must be used in proofs.

Theorem 5.3.2 $\frac{a}{b} \sim \frac{n \cdot a}{n \cdot b}$ *for any counting number n.*

Theorem 5.3.3 *The notion of equivalence has the following properties:*

1. *reflexive:* $\frac{a}{b} \sim \frac{a}{b}$,

2. *symmetric: if* $\frac{a}{b} \sim \frac{c}{d}$ *then* $\frac{c}{d} \sim \frac{a}{b}$,

3. *transitive: if* $\frac{a}{b} \sim \frac{c}{d}$ *and* $\frac{c}{d} \sim \frac{e}{f}$ *then* $\frac{a}{b} \sim \frac{e}{f}$.

Proof (of part 3)

Suppose $\frac{a}{b} \sim \frac{c}{d}$ and $\frac{c}{d} \sim \frac{e}{f}$. Then by definition, $a \cdot d = b \cdot c$ and $c \cdot f = d \cdot e$. From the first of these, $a \cdot d \cdot f = b \cdot c \cdot f$ and from the second $b \cdot c \cdot f = b \cdot d \cdot e$. Combining these equalities gives us $a \cdot d \cdot f = b \cdot d \cdot e$. Now $d \neq 0$ since it is on the bottom in $\frac{c}{d}$, so we can cancel it, getting $a \cdot f = b \cdot e$ which says $\frac{a}{b} \sim \frac{e}{f}$. $\square$

Exercises

Exercise 5.3.1 Which of the following are true?

1. $\frac{3}{4} \sim \frac{9}{12}$
2. $\frac{17}{32} \sim \frac{18}{32}$
3. $\frac{18}{32} \sim \frac{9}{16}$.

Exercise 5.3.2 Prove Theorem 5.3.2.

Exercise 5.3.3 Prove parts 1 and 2 of Theorem 5.3.3.

5.4   Rational numbers

Suppose line segment A is $\frac{3}{4}$ Units. That is

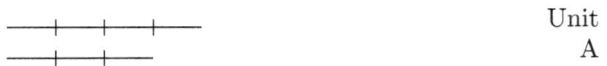

                                                                    Unit

                                                                     A

Then, as we saw in the last section, A is also $\frac{6}{8}$ Units, $\frac{9}{12}$ Unit, and so on. Each of these represents a different process by which A has been measured, but they all describe the same length. Any of $\frac{3}{4}$, $\frac{6}{8}$, $\frac{9}{12}$, etc. is equally correct as a description of the length of A, and none is more basic than any other. In a sense, it is the *collection* $\{\frac{3}{4}, \frac{6}{8}, \frac{9}{12}, \ldots\}$ that fully represents the length of A, since it embodies all the different special techniques for using our Unit to measure A. If it is *length*, rather than the *process of measurement* that we want to study, it is this collection we should work with. And so we give it, and other similar collections, names, and we develop an appropriate theory for them. We note that $\{\frac{3}{4}, \frac{6}{8}, \frac{9}{12}, \ldots\}$ is the collection consisting of precisely those fractions equivalent to $\frac{3}{4}$ (or to $\frac{6}{8}$, or to $\frac{9}{12}$, etc.). This suggests the following.

Definition 5.4.1 Let $\frac{a}{b}$ be a fraction. By the *rational number* corresponding to $\frac{a}{b}$ we mean the collection consisting of all those fractions equivalent to $\frac{a}{b}$. We write $\left[\frac{a}{b}\right]$ to denote this collection.

Remark The word 'rational' comes from the Latin word 'reri', and its past participle 'ratus'. The Latin word means to think, or estimate. One of its descendants is the term 'rational' for 'able to think.' In the Middle Ages the Latin word was commonly used to mean 'computation.' From it also comes our word 'ratio', which is closely related to 'rational' as we are using it.

Our definition says $\frac{x}{y} \in \left[\frac{a}{b}\right]$ precisely when $\frac{x}{y} \sim \frac{a}{b}$. For example, by Theorem 5.3.3, $\frac{a}{b} \sim \frac{a}{b}$, so $\frac{a}{b} \in \left[\frac{a}{b}\right]$. Similarly, by Theorem 5.3.2, $\frac{a}{b} \sim \frac{n \cdot a}{n \cdot b}$ so $\frac{a}{b} \in \left[\frac{n \cdot a}{n \cdot b}\right]$.

Continuing from earlier, $\left[\frac{3}{4}\right] = \{\frac{3}{4}, \frac{6}{8}, \frac{9}{12}, \ldots\}$ and $\left[\frac{6}{8}\right] = \{\frac{3}{4}, \frac{6}{8}, \frac{9}{12}, \ldots\}$ which means $\left[\frac{3}{4}\right] = \left[\frac{6}{8}\right]$. The following is a general statement of this phenomenon.

Theorem 5.4.2 $\frac{a}{b} \sim \frac{c}{d}$ *if and only if* $\left[\frac{a}{b}\right] = \left[\frac{c}{d}\right]$.

Remark In words this says equivalent fractions determine the same rational numbers, and conversely. This is as it should be since, intuitively, equivalent fractions represent measurements of the same line segment, and it is the length of this line segment that the rational number is intended to represent.

Proof Since the theorem is an 'if and only if' statement, the proof has two parts.

I. Suppose $\left[\frac{a}{b}\right] = \left[\frac{c}{d}\right]$. By Theorem 5.3.3, $\frac{a}{b} \sim \frac{a}{b}$, so $\frac{a}{b} \in \left[\frac{a}{b}\right]$. But $\left[\frac{a}{b}\right] = \left[\frac{c}{d}\right]$; they are the same collections. Since $\frac{a}{b} \in \left[\frac{a}{b}\right]$, we have $\frac{a}{b} \in \left[\frac{c}{d}\right]$. This means $\frac{a}{b} \sim \frac{c}{d}$.

II. Suppose $\frac{a}{b} \sim \frac{c}{d}$. We show $\left[\frac{a}{b}\right]$ and $\left[\frac{c}{d}\right]$ are the same collections, by showing they have the same things in them.

Suppose $\frac{x}{y} \in \left[\frac{a}{b}\right]$. We show $\frac{x}{y}$ is also in $\left[\frac{c}{d}\right]$. Well, since $\frac{x}{y} \in \left[\frac{a}{b}\right]$, $\frac{x}{y} \sim \frac{a}{b}$. We are given that $\frac{a}{b} \sim \frac{c}{d}$. By Theorem 5.3.3 (transitivity) $\frac{x}{y} \sim \frac{c}{d}$ and so $\frac{x}{y} \in \left[\frac{c}{d}\right]$.

We leave the rest to you as an exercise. □

Remark $\frac{1}{2} \neq \frac{2}{4}$. They are different fractions, and represent different measuring processes. $\frac{1}{2} \sim \frac{2}{4}$ since $1 \cdot 4 = 2 \cdot 2$. Then informally $\frac{1}{2}$ and $\frac{2}{4}$ are measurements of the same line segment. By the theorem above, $\left[\frac{1}{2}\right] = \left[\frac{2}{4}\right]$, the rational numbers which correspond to $\frac{1}{2}$ and $\frac{2}{4}$, are the same.

Exercises

Exercise 5.4.1 Complete this proof of Theorem 5.4.2 by showing that if $\frac{x}{y} \in \left[\frac{c}{d}\right]$ then $\frac{x}{y} \in \left[\frac{a}{b}\right]$.

Exercise 5.4.2 Which of the following are true:

1. $\left[\frac{3}{4}\right] = \left[\frac{9}{12}\right]$,

2. $\left[\frac{17}{32}\right] = \left[\frac{18}{32}\right]$,

3. $\left[\frac{18}{32}\right] = \left[\frac{9}{16}\right]$.

Exercise 5.4.3 The collection of all fractions with 0 on top is a rational number. Show this as follows. By definition, $\left[\frac{0}{1}\right]$ is a rational number. Show it is exactly the collection of fractions with 0 on top by showing:

1. if $\frac{x}{y} \in \left[\frac{0}{1}\right]$, then $x = 0$;

2. if $x = 0$ then $\frac{x}{y} \in \left[\frac{0}{1}\right]$.

Exercise 5.4.4 The collection of all fractions with top and bottom equal is a rational number. Show this as follows. By definition, $\left[\frac{1}{1}\right]$ is a rational number. Show it is exactly the collection of fractions with top and bottom equal by showing:

1. if $\frac{x}{y} \in \left[\frac{1}{1}\right]$, then $x = y$;

2. for any counting number x, $\frac{x}{x} \in \left[\frac{1}{1}\right]$.

5.5   Addition of fractions

Suppose we have two line segments, A and B, and suppose we have measured each of them separately. Say A is $\frac{a}{b}$ Units and B is $\frac{c}{d}$ Units. Now suppose we put A and B together.

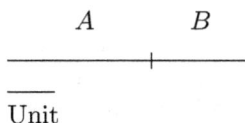

$$A \qquad\qquad B$$

———————————+————————

$\underline{}$
Unit

If we want the length of the combination, must we measure all over again, or is there a way of calculating it from $\frac{a}{b}$ and $\frac{c}{d}$?

We begin with a simple example, one for which fractions are not really needed. Say A is 3 Units and B is 2 Units.

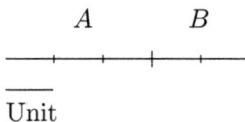

$$A \qquad\qquad B$$

———+———+———+———+————

$\underline{}$
Unit

Then it is clear that A and B are 5 Units, that is, $3 + 2$ Units.

Next, a slightly more complicated example. Say A is $\frac{3}{7}$ Unit and B is $\frac{2}{7}$ Unit.

———+——+—— A

———+—— B

———+——+——+——+——+——+——   Unit

Then again it is clear that the combination of A and B will be $3 + 2$ 7^{ths}, $\frac{3+2}{7}$ or $\frac{5}{7}$.

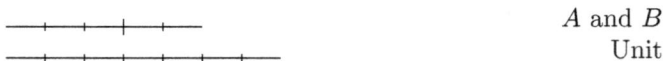

———+——+——+—— A and B

———+——+——+——+——+——   Unit

These two examples were easy since the standard of measurements for both A and B were the same, so all we had to do was add up the total number used.

Now suppose different standards have been used. Say A is $\frac{2}{9}$ Unit and B is $\frac{1}{4}$ Unit. (This time we have picked a bigger Unit, to make it easier to see what is happening.)

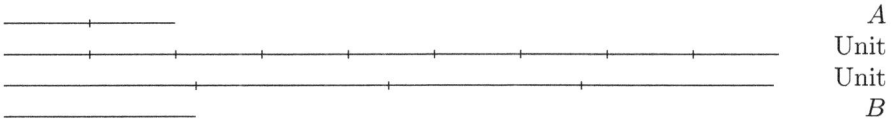

$$
\begin{array}{ll}
\rule{4cm}{0.4pt} & A \\
\rule{9cm}{0.4pt} & \text{Unit} \\
\rule{8cm}{0.4pt} & \text{Unit} \\
\rule{3cm}{0.4pt} & B
\end{array}
$$

But it is easy to change standards. If A is $\frac{2}{9}$ Units, A is also $\frac{2\cdot4}{9\cdot4}$ or $\frac{8}{36}$ Units. And if B is $\frac{1}{4}$ Unit, B is also $\frac{9\cdot1}{9\cdot4}$ or $\frac{9}{36}$ Unit. Now the standards are the same, namely 36^{ths}, and the total number used for A and B together is $8+9$ of them. So the combination of A with B should be $\frac{8+9}{36}$ or $\frac{17}{36}$ Unit.

Most generally, suppose A is $\frac{a}{b}$ Unit and B is $\frac{c}{d}$ Unit. Then also A is $\frac{a\cdot d}{b\cdot d}$ Unit and B is $\frac{b\cdot c}{b\cdot d}$ Unit. Then the total number of $(b\cdot d)^{ths}$ needed to make up A and B together is $a\cdot d + b\cdot c$, so A and B combined is $\frac{a\cdot d + b\cdot c}{b\cdot d}$ Unit.

The process of going from the fractions $\frac{a}{b}$ and $\frac{c}{d}$ to the fraction $\frac{a\cdot d + b\cdot c}{b\cdot d}$ is of interest to us because it corresponds to the combining of line segments. In our first example, where we did not use fractions, what we did was a whole number *addition*. Reasonably, we may call the more general operation, involving fractions, addition too.

Definition 5.5.1

$$
\frac{a}{b} + \frac{c}{d} = \frac{a\cdot d + b\cdot c}{b\cdot d}.
$$

Remark In using this definition it is simpler to learn the pattern rather than the formula.

1. The bottom is the product of the bottoms: $b\cdot d$;

2. The top is the sum of the 'cross products': $a\cdot d + b\cdot c$.

Exercises

Exercise 5.5.1 Use the definition and calculate:

1. $\frac{2}{9} + \frac{1}{4}$;

2. $\frac{1}{6} + \frac{2}{6}$.

5.6 Properties of addition of fractions

According to the definition,

$$\frac{1}{4} + \frac{2}{4} = \frac{1\cdot 4 + 4\cdot 2}{4\cdot 4} = \frac{12}{16}.$$

Now 1/4 and 2/4 have common denominators, so we are tempted to combine them into $\frac{1+2}{4}$ or $\frac{3}{4}$. As a matter of fact, $\frac{3}{4} \sim \frac{12}{16}$. The following theorem says this sort of thing always happens.

Theorem 5.6.1

$$\frac{a}{b} + \frac{c}{b} \sim \frac{a+c}{b}.$$

Proof By definition,

$$\frac{a}{b} + \frac{c}{b} = \frac{a\cdot b + b\cdot c}{b\cdot b}$$

$$= \frac{b\cdot(a+c)}{b\cdot b}$$

$$\sim \frac{a+c}{b} \qquad \text{by Theorem 5.3.2.}$$

□

According to the definition,

$$\frac{6}{15} + \frac{2}{10} = \frac{6\cdot 10 + 15\cdot 2}{15\cdot 10} = \frac{90}{150}.$$

But it is certainly tempting to say, $\frac{6}{15} = \frac{3\cdot 2}{3\cdot 5} \sim \frac{2}{5}$ and $\frac{2}{10} = \frac{2\cdot 1}{2\cdot 5} \sim \frac{1}{5}$ so instead of $\frac{6}{15} + \frac{2}{10}$ we may compute $\frac{2}{5} + \frac{1}{5}$, which, by Theorem 5.6.1, is equivalent to $\frac{3}{5}$. As a matter of fact, $\frac{3}{5} \sim \frac{90}{150}$. What we need is a general result that says we can always replace fractions by their equivalents in addition problems.

Theorem 5.6.2 *Suppose $\frac{a}{b} \sim \frac{a'}{b'}$ and $\frac{c}{d} \sim \frac{c'}{d'}$. Then $\frac{a}{b} + \frac{c}{d} \sim \frac{a'}{b'} + \frac{c'}{d'}$.*

Remark In words this says: sums of equivalent fractions are equivalent.

Proof We are given $\frac{a}{b} \sim \frac{a'}{b'}$ and $\frac{c}{d} \sim \frac{c'}{d'}$. By definition, this says $a\cdot b' = b\cdot a'$ (1) and $c\cdot d' = d\cdot c'$ (2). We must show $\frac{a}{b} + \frac{c}{d} \sim \frac{a'}{b'} + \frac{c'}{d'}$. By various definitions this says $(a\cdot d + b\cdot c)\cdot b'\cdot d' = b\cdot d\cdot(a'\cdot d' + b'\cdot c')$ or, using the distributive law, $a\cdot d\cdot b'\cdot d' + b\cdot c\cdot b'\cdot d' = b\cdot d\cdot a'\cdot d' + b\cdot d\cdot b'\cdot c'$ (3). The result will be established if we show (3) follows from (1) and (2), which is a whole number problem.

From (1), $a\cdot b'\cdot d\cdot d' = b\cdot a'\cdot d\cdot d'$ and from (2), $c\cdot d'\cdot b\cdot b' = d\cdot c'\cdot b\cdot b'$. Adding these together, $a\cdot b'\cdot d\cdot d' + c\cdot d'\cdot b\cdot b' = b\cdot a'\cdot d\cdot d' + d\cdot c'\cdot b\cdot b'$ and this is (3) except for the order of the factors, which we know doesn't matter. □

Theorem 5.6.3 (commutative law for addition)

$$\frac{a}{b} + \frac{c}{d} \sim \frac{c}{d} + \frac{a}{b}.$$

Theorem 5.6.4 (associative law for addition)

$$\frac{a}{b} + (\frac{c}{d} + \frac{e}{f}) \sim (\frac{a}{b} + \frac{c}{d}) + \frac{e}{f}.$$

Exercises

Exercise 5.6.1 Use Theorems 5.6.1 and 5.6.2 to show $\frac{3}{4} + \frac{7}{8} \sim \frac{13}{8}$.

Exercise 5.6.2 Prove Theorem 5.6.3.

Exercise 5.6.3 Prove Theorem 5.6.4.

5.7 Addition of rational numbers

Addition is designed to correspond to putting line segments together. We defined addition for fractions, but it is rational numbers that really represent lengths. We ought to have a notion of addition for rational numbers. Let us say the fraction $\frac{a}{b}$ *names* the rational number $[\frac{a}{b}]$. Reasonably, we would like to define addition for rational numbers in terms of addition of the fractions that name them. The problem is, rational numbers have many names; will it matter which one we choose? Let us look at an example. Suppose we try to 'add' $[\frac{2}{3}]$ and $[\frac{1}{2}]$ by working with names for them. Well, by definition of addition for fractions, $\frac{2}{3} + \frac{1}{2} = \frac{2\cdot2+3\cdot1}{3\cdot2} = \frac{7}{6}$ so we might set

$$\left[\frac{2}{3}\right] + \left[\frac{1}{2}\right] = \left[\frac{7}{6}\right].$$

But $[\frac{2}{3}] = [\frac{4}{6}]$ and $[\frac{1}{2}] = [\frac{2}{4}]$ since $\frac{2}{3} \sim \frac{4}{6}$ and $\frac{1}{2} \sim \frac{2}{4}$, so $\frac{4}{6}$ is also a name for $[\frac{2}{3}]$ and $\frac{2}{4}$ is also a name for $[\frac{1}{2}]$. If we work with these names we get $\frac{4}{6} + \frac{2}{4} = \frac{4\cdot4+6\cdot2}{6\cdot4} = \frac{28}{24}$ so we might also set

$$\left[\frac{2}{3}\right] + \left[\frac{1}{2}\right] = \left[\frac{28}{24}\right].$$

Fortunately, $\frac{28}{24} \sim \frac{7}{6}$, so $[\frac{28}{24}] = [\frac{7}{6}]$, we get the same thing either way.

The following theorem says this always happens, and makes possible our definition of addition for rational numbers.

Theorem 5.7.1 *Suppose* $\left[\frac{a}{b}\right] = \left[\frac{a'}{b'}\right]$ *and* $\left[\frac{c}{d}\right] = \left[\frac{c'}{d'}\right]$. *Then* $\left[\frac{a}{b} + \frac{c}{d}\right] = \left[\frac{a'}{b'} + \frac{c'}{d'}\right]$.

Proof This follows immediately from Theorems 5.4.2 and 5.6.2. □

Definition 5.7.2 $\left[\frac{a}{b}\right] + \left[\frac{c}{d}\right] = \left[\frac{a}{b} + \frac{c}{d}\right]$.

Remark This definition says we add rational numbers by adding their names, and seeing what rational number has the result in it. The theorem says it doesn't matter which names we choose to work with.

5.8 Properties of addition of rationals

We have, so far, used letters like a, b, c, ..., x, y, z to represent *whole numbers*. Now we start the convention of using capital letters, A, B, C, ..., X, Y, Z to represent *rational numbers*. Then if we say X is a rational number, we mean there are whole numbers x and y so that $X = \left[\frac{x}{y}\right]$. This convention of always using capital letters for rational numbers in this Chapter will save a great deal of writing.

Theorem 5.8.1 (commutativity of addition) $X + Y = Y + X$.

Proof X is a rational number, so $X = \left[\frac{a}{b}\right]$ for some fraction $\frac{a}{b}$. Likewise, $Y = \left[\frac{c}{d}\right]$ for some fraction $\frac{c}{d}$. Now, by Theorem 5.6.3, $\frac{a}{b} + \frac{c}{d} \sim \frac{c}{d} + \frac{a}{b}$. Then, by Theorem 5.4.2, $\left[\frac{a}{b} + \frac{c}{d}\right] = \left[\frac{c}{d} + \frac{a}{b}\right]$. By the definition of addition for rational numbers this says $\left[\frac{a}{b}\right] + \left[\frac{c}{d}\right] = \left[\frac{c}{d}\right] + \left[\frac{a}{b}\right]$, that is, $X + Y = Y + X$. □

Theorem 5.8.2 (associativity of addition) $X + (Y + Z) = (X + Y) + Z$.

Example Finally, we consider a specific example of addition, carried out rather as we are used to. $\left[\frac{1}{2}\right] + \left[\frac{3}{4}\right] = \left[\frac{1 \cdot 2}{2 \cdot 2}\right] + \left[\frac{3}{4}\right] = \left[\frac{2}{4}\right] + \left[\frac{3}{4}\right] = \left[\frac{2}{4} + \frac{3}{4}\right] = \left[\frac{5}{4}\right]$.

Exercises

Exercise 5.8.1 Prove Theorem 5.8.2.

Exercise 5.8.2 Justify each step of the example above by citing results we have proved, or definitions.

Exercise 5.8.3 Prove $X + \left[\frac{0}{1}\right] = X$.

5.9 Multiplication of fractions

We began our discussion of measuring by picking a standard Unit length. Well, suppose we choose a different standard, a *New Unit*, and now we want lengths in terms of New Units. Must we re-measure everything, or is it enough to measure our Old Unit, using our New Unit, and then do some calculation? More specifically, suppose we have measured line segment A using our Old

Unit, and also we have measured our Old Unit, using our New Unit. Can we *calculate* the length of A in terms of our New Unit, or must we go back and measure it all over?

We begin with a simple example, where fractions are not needed. Suppose A is 3 Old Units.

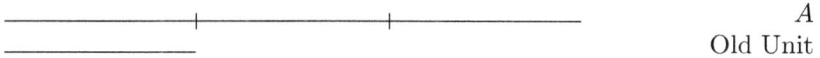

$$A$$
Old Unit

and suppose our Old Unit is 2 New Units

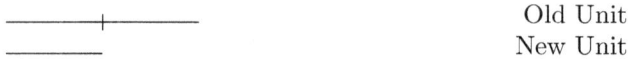

Old Unit
New Unit

Then there are 2 New Units in each Old Unit and 3 of those in A, so there are $2 \cdot 3 = 6$ New Units in A.

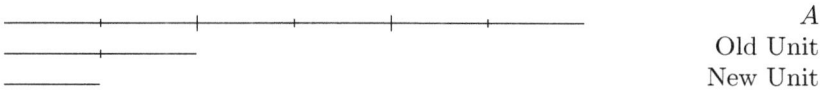

$$A$$
Old Unit
New Unit

For our next, more complicated, example, suppose line segment A is $\frac{2}{7}$ Old Unit, and our Old Unit is $\frac{7}{9}$ New Unit. Then the situation is still rather simple. We know the Old Unit is $\frac{7}{9}$ New Unit. So if we divide our New Unit into 9 parts, 7 of them will make up our Old Unit. But then each of these parts is a 7th of the Old Unit; we know A is $\frac{2}{7}$ Old Unit, that is, A is 2 of these parts, each of which is a 9th of the New Unit.

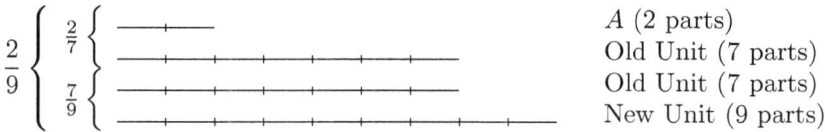

$$\frac{2}{9}\left\{ \begin{array}{l} \frac{2}{7}\left\{ \rule{0pt}{0pt} \right. \\ \frac{7}{9}\left\{ \rule{0pt}{0pt} \right. \end{array} \right.$$

A (2 parts)
Old Unit (7 parts)
Old Unit (7 parts)
New Unit (9 parts)

Thus A is $\frac{2}{9}$ New Units. The important thing in this example is that the bottom number in $\frac{2}{7}$ is also the top number in $\frac{7}{9}$.

More generally suppose

A is $\frac{a}{b}$ Old Unit, and
Old Unit is $\frac{b}{c}$ New Unit.

Since Old Unit is $\frac{b}{c}$ New Unit we can divide our New Unit into c parts, and use b of them to make up our Old Unit. Also, since A is $\frac{a}{b}$ Old Unit, we can divide our Old Unit up into b parts, and use a of them to make up A. But our Old Unit is already divided into b parts!

$$\frac{a}{c}\left\{\begin{array}{l}\frac{a}{b}\left\{\rule{0pt}{0pt}\right.\\\left.\rule{0pt}{0pt}\right.\\\frac{b}{c}\left\{\rule{0pt}{0pt}\right.\end{array}\right.$$

A (a parts)
Old Unit (b parts)
Old Unit (b parts)
New Unit (c parts)

In this diagram all the parts are the same size. It follows that A is $\frac{a}{c}$ New Unit.

Thus we know what to do if the bottom number in our measurement of A, $\frac{a}{b}$, is also the top number in our measurement of the Old Unit, $\frac{b}{c}$. We go from $\frac{a}{b}$ and $\frac{b}{c}$ to $\frac{a}{c}$.

Now consider the general case. Suppose A is $\frac{a}{b}$ Old Unit, and Old Unit is $\frac{c}{d}$ New Unit. Here the appropriate bottom and top numbers are not the same. But we can make them so. We have $\frac{a}{b} \sim \frac{a \cdot c}{b \cdot c}$ and $\frac{c}{d} \sim \frac{b \cdot c}{b \cdot d}$ so also A is $\frac{a \cdot c}{b \cdot c}$ Old Unit, and Old Unit is $\frac{b \cdot c}{b \cdot d}$ New Unit. Now the bottom number in our measurement of A is the same as the top number in our measurement of the Old Unit. By the argument above, then, we should go from $\frac{a \cdot c}{b \cdot c}$ and $\frac{b \cdot c}{b \cdot d}$ to $\frac{a \cdot c}{b \cdot d}$. The length of A should be $\frac{a \cdot c}{b \cdot d}$ New Unit.

The process of going from the fractions $\frac{a}{b}$ and $\frac{c}{d}$ to the fraction $\frac{a \cdot c}{b \cdot d}$ is of interest to us because it corresponds to changing our standard of measurement. In the first example, where fractions were not used, what we did was a whole number *multiplication*. We may, then, call the more general operation on fractions multiplication too.

Definition 5.9.1

$$\frac{a}{b} \cdot \frac{c}{d} = \frac{a \cdot c}{b \cdot d}.$$

Exercises

Exercise 5.9.1 In the example used above, $b \neq 0$ since it is on the bottom in $\frac{a}{b}$. Then $\frac{b \cdot c}{b \cdot d}$ is a legitimate expression. But it is possible for c to be 0. If it is, $\frac{a \cdot c}{b \cdot c}$ would not be a fraction as it would have 0 on the bottom. Our informal argument above tacitly assumed this did not happen. Now suppose A is $\frac{a}{b}$ Old Unit, and Old Unit is $\frac{0}{d}$ new Unit. See what this means, and give an informal argument that, even in this case, A is $\frac{a \cdot 0}{b \cdot d}$ New Unit.

Exercise 5.9.2 Show:

1. $\frac{3}{7} \cdot \frac{8}{11} = \frac{24}{77}$;

2. $\frac{3}{7} \cdot \frac{7}{3} \sim \frac{1}{1}$;

3. $\frac{3}{7} \cdot \frac{4}{4} \sim \frac{3}{7}$.

5.10 Properties of multiplication of fractions

Theorem 5.10.1 *Suppose $\frac{a}{b} \sim \frac{a'}{b'}$ and $\frac{c}{d} \sim \frac{c'}{d'}$. Then $\frac{a}{b} \cdot \frac{c}{d} \sim \frac{a'}{b'} \cdot \frac{c'}{d'}$.*

Remark This says that products of equivalent fractions are equivalent.

Theorem 5.10.2 (Commutativity of multiplication)

$$\frac{a}{b} \cdot \frac{c}{d} \sim \frac{c}{d} \cdot \frac{a}{b}.$$

Theorem 5.10.3 (Associativity of multiplication)

$$\frac{a}{b} \cdot \left(\frac{c}{d} \cdot \frac{e}{f} \right) \sim \left(\frac{a}{b} \cdot \frac{c}{d} \right) \cdot \frac{e}{f}.$$

Theorem 5.10.4 (Distributive law)

$$\frac{a}{b} \cdot \left(\frac{c}{d} + \frac{e}{f} \right) \sim \left(\frac{a}{b} \cdot \frac{c}{d} \right) + \left(\frac{a}{b} \cdot \frac{e}{f} \right).$$

Exercises

Exercise 5.10.1 Prove Theorem 5.10.1.

Exercise 5.10.2 Prove Theorem 5.10.2.

Exercise 5.10.3 Prove Theorem 5.10.3.

Exercise 5.10.4 Prove Theorem 5.10.4.

5.11 Multiplication of rational numbers

Theorem 5.11.1 *Suppose $\left[\frac{a}{b} \right] = \left[\frac{a'}{b'} \right]$ and $\left[\frac{c}{d} \right] = \left[\frac{c'}{d'} \right]$. Then $\left[\frac{a}{b} \cdot \frac{c}{d} \right] = \left[\frac{a'}{b'} \cdot \frac{c'}{d'} \right]$.*

Proof This is immediate by Theorems 5.4.2 and 5.10.1. □

Definition 5.11.2

$$\left[\frac{a}{b} \right] \cdot \left[\frac{c}{d} \right] = \left[\frac{a}{b} \cdot \frac{c}{d} \right].$$

Remark This definition says we multiply rational numbers by multiplying their fraction names. The theorem above says it won't matter which names we choose to work with.

Theorem 5.11.3 (Commutativity of multiplication) $X \cdot Y = Y \cdot X$.

Theorem 5.11.4 (Associativity of multiplication) $X \cdot (Y \cdot Z) = (X \cdot Y) \cdot Z$.

Theorem 5.11.5 (Distributive law) $X \cdot (Y + Z) = X \cdot Y + X \cdot Z$.

Exercises

Exercise 5.11.1 Prove Theorem 5.11.3.

Exercise 5.11.2 Prove Theorem 5.11.4.

Exercise 5.11.3 Prove Theorem 5.11.5.

Exercise 5.11.4 Show $X \cdot \left[\frac{0}{1}\right] = \left[\frac{0}{1}\right]$.

Exercise 5.11.5 Show $X \cdot \left[\frac{1}{1}\right] = X$.

Exercise 5.11.6 Show $\left[\frac{3}{7}\right] \cdot \left[\frac{7}{3}\right] = \left[\frac{1}{1}\right]$.

Exercise 5.11.7 Show $\left[\frac{2}{1}\right] \cdot X = X + X$.

5.12   Order of fractions

Suppose we use our Unit and measure each of two line segments, A and B.
Say A is $\frac{a}{b}$ Unit and B is $\frac{c}{d}$ Unit. Suppose A is longer than B. Then it is
reasonable to say $\frac{a}{b}$ is a bigger fraction than $\frac{c}{d}$. Now, is there some way we can
recognize when one fraction is bigger than another, without having to go out
and measure something?

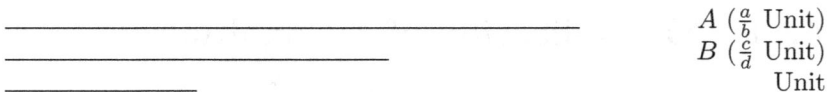

$$\begin{array}{ll} \rule{6cm}{0.4pt} & A \ (\frac{a}{b} \text{ Unit}) \\ \rule{4.5cm}{0.4pt} & B \ (\frac{c}{d} \text{ Unit}) \\ \rule{2.5cm}{0.4pt} & \text{Unit} \end{array}$$

If A is $\frac{a}{b}$ Unit, it is also $\frac{a \cdot d}{b \cdot d}$ Unit. Likewise if B is $\frac{c}{d}$ Unit, it is also $\frac{b \cdot c}{b \cdot d}$ Unit.

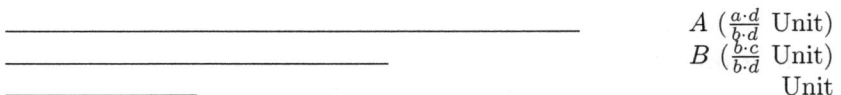

$$\begin{array}{ll} \rule{6cm}{0.4pt} & A \ (\frac{a \cdot d}{b \cdot d} \text{ Unit}) \\ \rule{4.5cm}{0.4pt} & B \ (\frac{b \cdot c}{b \cdot d} \text{ Unit}) \\ \rule{2.5cm}{0.4pt} & \text{Unit} \end{array}$$

Then, if we divide our Unit into $b \cdot d$ parts, we will need $a \cdot d$ of them to
make up A, and $b \cdot c$ of them to make up B. But we are supposing A is longer
than B, so we should need more parts to fill out A than to fill out B. We
should have $a \cdot d > b \cdot c$.

Definition 5.12.1 We say $\frac{a}{b} > \frac{c}{d}$ if $a \cdot d > b \cdot c$.

Definition 5.12.2 $\frac{a}{b} < \frac{c}{d}$ if $\frac{c}{d} > \frac{a}{b}$.

According to the definition, $\frac{3}{4} > \frac{2}{3}$ since $3 \cdot 3 > 4 \cdot 2$. But $\frac{3}{4} \sim \frac{6}{8}$, that is, if $\frac{3}{4}$ comes up when we measure a certain line segment, $\frac{6}{8}$ can also come up. Since $>$ for fractions was meant to relate lengths of line segments, we ought to have $\frac{6}{8} > \frac{2}{3}$ also, since we had $\frac{3}{4} > \frac{2}{3}$. As a matter of fact, we do have $\frac{6}{8} > \frac{2}{3}$ since $6 \cdot 3 > 8 \cdot 2$. The following theorem says this will always happen.

Theorem 5.12.3 *Suppose $\frac{a}{b} \sim \frac{a'}{b'}$ and $\frac{c}{d} \sim \frac{c'}{d'}$. Then if $\frac{a}{b} > \frac{c}{d}$ then $\frac{a'}{b'} > \frac{c'}{d'}$.*

Proof Using the definitions of $\sim$ and $>$ for fractions, we are given

$a \cdot b' = b \cdot a'$

$c \cdot d' = d \cdot c'$

$a \cdot d > b \cdot c$

and we must show

$a' \cdot d' > b' \cdot c'$.

This is a problem about whole numbers. Now,

$a \cdot d > b \cdot c$

Also, neither b' nor d' is 0, so we may use Theorem 3.5.3 to conclude

$a \cdot d \cdot b' \cdot d' > b \cdot c \cdot b' \cdot d'$

or, rearranging,

$a \cdot b' \cdot d \cdot d' > b \cdot b' \cdot c \cdot d'$.

Next, $a \cdot b' = b \cdot a'$ and $c \cdot d' = d \cdot c'$, so on substituting, we get

$b \cdot a' \cdot d \cdot d' > b \cdot b' \cdot d \cdot c'$

Next, neither b nor d is 0, so by using Theorem 3.5.3 again,

$a' \cdot d' > b' \cdot c'$

which completes the proof. □

Theorem 5.12.4 (Transitivity of $>$) *If $\frac{a}{b} > \frac{c}{d}$ and $\frac{c}{d} > \frac{e}{f}$ then $\frac{a}{b} > \frac{e}{f}$.*

Proof Suppose $\frac{a}{b} > \frac{c}{d}$ and $\frac{c}{d} > \frac{e}{f}$. Then by definition,

1) $a \cdot d > b \cdot c$
2) $c \cdot f > d \cdot e$

Since f is on the bottom in $\frac{e}{f}$, $f \neq 0$, so from *1)*,

$a \cdot d \cdot f > b \cdot c \cdot f$.

Similarly $b \neq 0$ since $\frac{a}{b}$ is a fraction, so from *2)*,

$b \cdot c \cdot f > b \cdot d \cdot e$.

Now by transitivity of $>$ for whole numbers (Theorem 3.4.4),

$a \cdot d \cdot f > b \cdot d \cdot e$.

Finally, $d \neq 0$, since $\frac{c}{d}$ is a fraction, so from this,

$a \cdot f > b \cdot e$

which means

$\frac{a}{b} > \frac{e}{f}$. □

Theorem 5.12.5 (Trichotomy law) *For each choice of fractions $\frac{a}{b}$ and $\frac{c}{d}$ we have exactly one of the following:*

    1. $\frac{a}{b} > \frac{c}{d}$,

    2. $\frac{a}{b} \sim \frac{c}{d}$,

    3. $\frac{a}{b} < \frac{c}{d}$.

Definition 5.12.6 $\frac{a}{b} \geq \frac{c}{d}$ if $\frac{a}{b} > \frac{c}{d}$ or $\frac{a}{b} \sim \frac{c}{d}$.

Exercises

Exercise 5.12.1 Prove Theorem 5.12.5.

Exercise 5.12.2 Prove $\frac{a}{b} \geq \frac{c}{d}$ if and only if $a \cdot d \geq b \cdot c$.

5.13   Order of rational numbers

Theorem 5.13.1 *Suppose $\left[\frac{a}{b}\right] = \left[\frac{a'}{b'}\right]$ and $\left[\frac{c}{d}\right] = \left[\frac{c'}{d'}\right]$. Then $\frac{a}{b} > \frac{c}{d}$ if and only if $\frac{a'}{b'} > \frac{c'}{d'}$.*

Proof Immediate from Theorems 5.4.2 and 5.12.3. □

Definition 5.13.2 $\left[\frac{a}{b}\right] > \left[\frac{c}{d}\right]$ if $\frac{a}{b} > \frac{c}{d}$.

Remark As expected, this definition has us compare rational numbers by comparing their fraction names. The theorem above says our choice of names is not important.

Theorem 5.13.3 (Transitivity of >) *If $X > Y$ and $Y > Z$ then $X > Z$.*

Theorem 5.13.4 (Trichotomy law) *For any rational numbers X and Y, exactly one of the following holds: $X > Y$, $X = Y$, $X < Y$.*

Definition 5.13.5 $X \geq Y$ if $X > Y$ or $X = Y$.

Both the whole numbers and the rational numbers have a notion of order. And these two notions have many similar properties; both are transitive, for example. But they also have important differences. We saw in Chapter Three that there were no whole numbers between n and n^+. But in the system of rational numbers, between any two there is another. We leave this to you to verify, as Exercise 5.13.5.

Exercises

Exercise 5.13.1 Prove Theorem 5.13.3.

Exercise 5.13.2 Prove Theorem 5.13.4.

Exercise 5.13.3 Prove $\left[\frac{a}{b}\right] \geq \left[\frac{c}{d}\right]$ if and only if $\frac{a}{b} \geq \frac{c}{d}$.

Exercise 5.13.4 Prove $\left[\frac{a}{b}\right] \geq \left[\frac{c}{d}\right]$ if and only if $a \cdot d \geq b \cdot c$.

Exercise 5.13.5 Suppose $A > B$. Show $A > \left[\frac{1}{2}\right] \cdot (A+B)$ and $\left[\frac{1}{2}\right] \cdot (A+B) > B$.

Exercise 5.13.6 Show that between any two rational numbers there is an unlimited number of others.

5.14 Insertion and deletion laws

The pattern of our work in this chapter has been to prove some result for fractions, then use it to get a similar result for rational numbers. In this section we only state the results for rational numbers, but, of course, we work with fractions to establish them.

Theorem 5.14.1 *For any rational numbers X, Y and Z, $X = Y$ if and only if $X + Z = Y + Z$.*

Proof If $X = Y$ it is immediate that $X + Z = Y + Z$. Now suppose $X + Z = Y + Z$. We show $X = Y$. X, Y and Z are rational numbers, say $X = \left[\frac{x}{y}\right]$, $Y = \left[\frac{z}{w}\right]$ and $Z = \left[\frac{a}{b}\right]$. Thus we have:

$$\left[\frac{x}{y}\right] + \left[\frac{a}{b}\right] = \left[\frac{z}{w}\right] + \left[\frac{a}{b}\right]$$

$$\left[\frac{x}{y} + \frac{a}{b}\right] = \left[\frac{z}{w} + \frac{a}{b}\right]$$

$$\left[\frac{x \cdot b + y \cdot a}{y \cdot b}\right] = \left[\frac{z \cdot b + w \cdot a}{w \cdot b}\right]$$

$$\frac{x \cdot b + y \cdot a}{y \cdot b} \sim \frac{z \cdot b + w \cdot a}{w \cdot b}$$

$$(x \cdot b + y \cdot a) \cdot w \cdot b = y \cdot b \cdot (z \cdot b + w \cdot a).$$

Now $b \neq 0$ since it is on the bottom in $\frac{a}{b}$ so we may cancel it.

$$(x \cdot b + y \cdot a) \cdot w = y \cdot (z \cdot b + w \cdot a)$$

$$x \cdot b \cdot w + y \cdot a \cdot w = y \cdot z \cdot b + y \cdot w \cdot a.$$

Next, $y \cdot a \cdot w$ occurs on both sides; cancelling it, we get:

$$x \cdot b \cdot w = y \cdot z \cdot b.$$

Again, $b \neq 0$ so:

$$x \cdot w = y \cdot z$$

$$\frac{x}{y} \sim \frac{z}{w}$$

$$\left[\frac{x}{y} \right] = \left[\frac{z}{w} \right]$$

$$X = Y.$$

This concludes the proof. $\square$

Theorem 5.14.2 *For any rational numbers X, Y and Z, $X > Y$ if and only if $X + Z > Y + Z$.*

Theorem 5.14.3 *For any rational numbers X, Y and Z, with $Z \neq \left[\frac{0}{1} \right]$:*

1. *$X = Y$ if and only if $X \cdot Z = Y \cdot Z$;*

2. *$X > Y$ if and only if $X \cdot Z > Y \cdot Z$.*

Exercises

Exercise 5.14.1 Prove Theorem 5.14.2.

Exercise 5.14.2 Prove Theorem 5.14.3.

5.15 Subtraction

We will define subtraction for rational numbers directly, but to calculate particular subtractions, we will have to work with fractions.

We want to define subtraction so that it is 'opposite to' addition. Recall, when we worked with whole numbers, $x - y$ was that unique whole number which, when added to y, gave x. But, in order to use that definition we had to determine two things:

1. when would there be a number we could add to y to get x;

2. when would there be only one.

We face the same problems in working with rational numbers.

Theorem 5.15.1 *If $Y + A = X$ and also $Y + A' = X$ then $A = A'$.*

Proof $Y + A = X$ and $Y + A' = X$ so $Y + A = Y + A'$. Now by Theorem 5.14.1, $A = A'$. $\square$

This theorem says that if there is anything we can add to Y to get X, there is only one thing. Now we must determine when such a thing will exist. Reasonably, we begin by looking at some *fraction* examples. In them, of course, we do not try for true equality; equivalence is enough.

Consider $\frac{3}{7}$ and $\frac{5}{7}$. Is there some fraction $\frac{a}{b}$ such that $\frac{3}{7} + \frac{a}{b} \sim \frac{5}{7}$? Since both $\frac{3}{7}$ and $\frac{5}{7}$ have 7 on the bottom, it is reasonable to try $b = 7$. So we ask, is there an a for which $\frac{3}{7} + \frac{a}{7} \sim \frac{5}{7}$. By Theorem 5.6.1, $\frac{3}{7} + \frac{a}{7} \sim \frac{3+a}{7}$, so the question becomes: is there an a for which $\frac{3+a}{7} \sim \frac{5}{7}$? To do this we simply need an a for which $3 + a = 5$. And by our work with whole numbers, we know $a = 5 - 3 = 2$. Thus $\frac{3}{7} + \frac{2}{7} \sim \frac{5}{7}$.

Next, consider $\frac{1}{9}$ and $\frac{5}{7}$. Is there a fraction $\frac{a}{b}$ such that $\frac{1}{9} + \frac{a}{b} \sim \frac{5}{7}$? Here the denominators are different, but we can make them the same since $\frac{1}{9} \sim \frac{1\cdot7}{9\cdot7} = \frac{7}{63}$ and $\frac{5}{7} \sim \frac{5\cdot9}{7\cdot9} = \frac{45}{63}$. Then our question becomes: is there a fraction $\frac{a}{b}$ for which $\frac{7}{63} + \frac{a}{b} \sim \frac{45}{63}$, and following the pattern of our previous example, we come up with $\frac{45-7}{63}$, or $\frac{7}{63} + \frac{38}{63} \sim \frac{45}{63}$.

More generally, consider $\frac{x}{y}$ and $\frac{z}{w}$. Is there a fraction $\frac{a}{b}$ for which $\frac{z}{w} + \frac{a}{b} \sim \frac{x}{y}$? As in the previous example, we introduce common denominators; $\frac{x}{y} \sim \frac{x\cdot w}{y\cdot w}$ and $\frac{z}{w} \sim \frac{y\cdot z}{y\cdot w}$, so we are asking, is there a fraction $\frac{a}{b}$ for which $\frac{y\cdot z}{y\cdot w} + \frac{a}{b} \sim \frac{x\cdot w}{y\cdot w}$? Now, our previous example suggests we should try $\frac{a}{b} = \frac{x\cdot w - y\cdot z}{y\cdot w}$. But notice, this only makes sense if $x \cdot w - y \cdot z$ is defined, which happens when $x \cdot w \geq y \cdot z$. And by Exercise 5.12.2 this happens just when $\frac{x}{y} \geq \frac{z}{w}$.

Lemma 5.15.2 *If $\frac{x}{y} \geq \frac{z}{w}$ then $\frac{x\cdot w - y\cdot z}{y\cdot w}$ is defined, and $\frac{z}{w} + \frac{x\cdot w - y\cdot z}{y\cdot w} \sim \frac{x}{y}$.*

Now we return to rational numbers.

Theorem 5.15.3 *The following two items are equivalent:*

1. for some A, Y + A = X

2. X ≥ Y.

Proof I. Suppose first that for some A, $Y + A = X$. X, Y and A are rational numbers, say $X = \left[\dfrac{x}{y}\right]$, $Y = \left[\dfrac{z}{w}\right]$ and $A = \left[\dfrac{a}{b}\right]$. Then

$$\left[\frac{z}{w}\right] + \left[\frac{a}{b}\right] = \left[\frac{x}{y}\right]$$

$$\left[\frac{z}{w} + \frac{a}{b}\right] = \left[\frac{x}{y}\right]$$

$$\left[\frac{z \cdot b + w \cdot a}{w \cdot b}\right] = \left[\frac{x}{y}\right]$$

$$\frac{z \cdot b + w \cdot a}{w \cdot b} \sim \frac{x}{y}$$

$$(z \cdot b + w \cdot a) \cdot y = x \cdot w \cdot b$$

$$z \cdot b \cdot y + w \cdot a \cdot y = x \cdot w \cdot b$$

But by the properties of whole numbers, this says,

$$x \cdot w \cdot b \geq z \cdot b \cdot y.$$

Now $b \neq 0$ since it is on the bottom in $\frac{a}{b}$, so we can cancel it.

$$x \cdot w \geq z \cdot y$$

and by Exercise 5.13.4, this says,

$$\left[\frac{x}{y}\right] \geq \left[\frac{z}{w}\right]$$

or

$$X \geq Y.$$

II. Now suppose $X \geq Y$. Again X and Y are rational numbers, say $X = \begin{bmatrix} \dfrac{x}{y} \end{bmatrix}$
and $Y = \begin{bmatrix} \dfrac{z}{w} \end{bmatrix}$. Then

$$\begin{bmatrix} \frac{x}{y} \end{bmatrix} \geq \begin{bmatrix} \frac{z}{w} \end{bmatrix}.$$

By Exercise 5.13.3,

$$\frac{x}{y} \geq \frac{z}{w}.$$

Then by Lemma 5.15.2, $\frac{x \cdot w - y \, z}{y \cdot w}$ is meaningful. We set

$$A = \begin{bmatrix} \dfrac{x \cdot w - y \cdot z}{y \cdot w} \end{bmatrix}$$

and then Lemma 5.15.2 immediately gives us

$$Y + A = X.$$

$\square$

Now we know when there will be something we can add to Y to get X; it happens when $X \geq Y$.

Definition 5.15.4 If $X \geq Y$ then $X - Y$ is defined, and it is that unique A for which $Y + A = X$.

As far as actually *calculating* subtractions, we have the following.

Theorem 5.15.5

1. If $\begin{bmatrix} \dfrac{x}{y} \end{bmatrix} \geq \begin{bmatrix} \dfrac{z}{w} \end{bmatrix}$ then $\begin{bmatrix} \dfrac{x}{y} \end{bmatrix} - \begin{bmatrix} \dfrac{z}{w} \end{bmatrix} = \begin{bmatrix} \dfrac{x \cdot w - y \cdot z}{y \cdot w} \end{bmatrix}.$

2. If $\begin{bmatrix} \dfrac{x}{w} \end{bmatrix} \geq \begin{bmatrix} \dfrac{z}{w} \end{bmatrix}$ then $\begin{bmatrix} \dfrac{x}{w} \end{bmatrix} - \begin{bmatrix} \dfrac{z}{w} \end{bmatrix} = \begin{bmatrix} \dfrac{x - z}{w} \end{bmatrix}.$

Exercises

Exercise 5.15.1 Prove Lemma 5.15.2.

Exercise 5.15.2 Prove Theorem 5.15.5.

Exercise 5.15.3 Justify each step in the following. $\begin{bmatrix} \frac{5}{6} \end{bmatrix} - \begin{bmatrix} \frac{1}{3} \end{bmatrix} = \begin{bmatrix} \frac{5}{6} \end{bmatrix} - \begin{bmatrix} \frac{2}{6} \end{bmatrix} = \begin{bmatrix} \frac{5-2}{6} \end{bmatrix} = \begin{bmatrix} \frac{3}{6} \end{bmatrix} = \begin{bmatrix} \frac{1}{2} \end{bmatrix}.$

Exercise 5.15.4 Prove $X - X = \begin{bmatrix} \frac{0}{1} \end{bmatrix}.$

Exercise 5.15.5 Prove $X \geq \begin{bmatrix} \frac{0}{1} \end{bmatrix}$ for any X, and $X - \begin{bmatrix} \frac{0}{1} \end{bmatrix} = X.$

5.16 Properties of subtraction

The basic properties of subtraction for rational numbers are the same as for
whole numbers. We list some of them in the theorem below. To cut down on
words, we will simply write, say, $X - Y = (X - Z) + (Z - Y)$ and mean by it:
if $X - Y$, $X - Z$ and $Z - Y$ are all defined, then $X - Y = (X - Z) + (Z - Y)$.

Theorem 5.16.1

1. $X - Y = (X - Z) + (Z - Y)$.

2. $X \cdot (Y - Z) = X \cdot Y - X \cdot Z$.

3. $(X + Y) - Z = X + (Y - Z)$.

4. $X - (Y + Z) = (X - Y) - Z = (X - Z) - Y$.

5. $X - Y = (X + Z) - (Y + Z)$.

6. $X - (Y - Z) = (X + Z) - Y$.

7. $X - (Y - Z) = (X - Y) + Z$.

Exercises

Exercise 5.16.1 Prove Theorem 5.16.1.

5.17 Division

We want an operation to go opposite to multiplication of rational numbers.
For whole numbers we said x was divisible by y if there was exactly one z such
that $y \cdot z = x$, and if there was, we set $x \div y$ to be that z. We follow the same
pattern for rational numbers. We must examine the questions: when will there
be some Z for which $Y \cdot Z = X$; when will it be unique?

Theorem 5.17.1 *If $Y \cdot A = X$ and $Y \cdot A' = X$, where $Y \neq \left[\frac{0}{1}\right]$, then $A = A'$.*

Proof $Y \cdot A = X$ and $Y \cdot A' = X$, so $Y \cdot A = Y \cdot A'$. Now by Theorem 5.14.3,
$A = A'$. □

This theorem says that, provided $Y \neq \left[\frac{0}{1}\right]$, if there is anything by which
we can multiply Y to get X, there is only one thing. But when will there be
anything? We begin by looking at *fraction* examples. And, of course, we only
require equivalence, not equality.

Is there some fraction $\frac{a}{b}$ such that $\frac{2}{5} \cdot \frac{a}{b} \sim \frac{3}{4}$? Suppose there is, let us
find one. Using the definition of multiplication, we would have $\frac{2 \cdot a}{5 \cdot b} \sim \frac{3}{4}$. By

definition $2 \cdot a \cdot 4 = 5 \cdot b \cdot 3$. Rearranging this, $a \cdot 2 \cdot 4 = b \cdot 5 \cdot 3$ which says $\frac{a}{b} \sim \frac{5 \cdot 3}{2 \cdot 4}$.

More generally, consider $\frac{x}{y}$ and $\frac{z}{w}$. Is there a fraction, $\frac{a}{b}$, for which $\frac{z}{w} \cdot \frac{a}{b} \sim \frac{x}{y}$? If there were, by definition of multiplication, $\frac{z \cdot a}{w \cdot b} \sim \frac{x}{y}$ which gives $z \cdot a \cdot y = w \cdot b \cdot x$. Rearranging $a \cdot y \cdot z = b \cdot x \cdot w$ which gives $\frac{a}{b} \sim \frac{x \cdot w}{y \cdot z}$. There is a problem hidden here, however. We know $y \neq 0$ since it is on the bottom in $\frac{x}{y}$. But z is on the top in $\frac{z}{w}$, so z could be 0. If it is, $\frac{x \cdot w}{y \cdot z}$ has 0 on the bottom, and so is not a fraction. The above only works if $z \neq 0$.

Theorem 5.17.2 *If $Y \neq \left[\frac{0}{1}\right]$ then there is an A for which $Y \cdot A = X$.*

Proof X and Y are rational numbers, say $X = \left[\frac{x}{y}\right]$ and $Y = \left[\frac{z}{w}\right]$. Since $Y \neq \left[\frac{0}{1}\right]$, $z \neq 0$ (why?). Then $\frac{x \cdot w}{y \cdot z}$ is a fraction. We set $A = \left[\frac{x \cdot w}{y \cdot z}\right]$. Then Exercise 5.17.2 gives us that $Y \cdot A = X$. □

Now we know when there will be something by which we can multiply Y to get X; it happens when $Y \neq \left[\frac{0}{1}\right]$.

Definition 5.17.3 Suppose $Y \neq \left[\frac{0}{1}\right]$. Then $X \div Y$ is that unique rational number A such that $Y \cdot A = X$.

Notice that we don't need anything comparable to divison with remainder to handle the cases where exact division isn't possible. In the rational number system, exact division, except by $\left[\frac{0}{1}\right]$, always happens.

Now we look at how one might *calculate* divisions. In Exercise 5.17.2 we see $\frac{z}{w} \cdot \frac{x \cdot w}{y \cdot z} \sim \frac{x}{y}$ which gives $\left[\frac{z}{w}\right] \cdot \left[\frac{x \cdot w}{y \cdot z}\right] = \left[\frac{x}{y}\right]$ and this means that $\left[\frac{x}{y}\right] \div \left[\frac{z}{w}\right] = \left[\frac{x \cdot w}{y \cdot z}\right]$. As it turns out, there is an easy way to describe this rational number.

$\left[\frac{x \cdot w}{y \cdot z}\right] = \left[\frac{x}{y} \cdot \frac{w}{z}\right] = \left[\frac{x}{y}\right] \cdot \left[\frac{w}{z}\right]$. Here $\left[\frac{x}{y}\right]$ is one of the rational numbers we

started with, and $\left[\frac{w}{z}\right]$ is the other one 'upside down'.

Definition 5.17.4 If $\left[\frac{a}{b}\right] \neq \left[\frac{0}{1}\right]$ then $\left[\frac{a}{b}\right]^{-1} = \left[\frac{b}{a}\right]$.

Remark If $\left[\frac{a}{b}\right] \neq \left[\frac{0}{1}\right]$ then no fraction in $\left[\frac{a}{b}\right]$ can have 0 on top (Why?). So turning any member of $\left[\frac{a}{b}\right]$ over produces a fraction. Exercise 5.17.3 says it doesn't matter which member of $\left[\frac{a}{b}\right]$ we choose to turn over. The expression X^{-1} is read 'X inverse'.

Using this notion of inverse, the rational number we found earlier becomes simply $\left[\frac{x \cdot w}{y \cdot z}\right] = \left[\frac{x}{y}\right] \cdot \left[\frac{z}{w}\right]^{-1}$. That is, to divide $\left[\frac{x}{y}\right]$ by $\left[\frac{z}{w}\right]$, we invert $\left[\frac{z}{w}\right]$ and multiply.

Exercises

Exercise 5.17.1 Show $\frac{2}{5} \cdot \frac{5 \cdot 3}{2 \cdot 4} \sim \frac{3}{4}$.

Exercise 5.17.2 Suppose $z \neq 0$. Show $\frac{z}{w} \cdot \frac{x \cdot w}{y \cdot z} \sim \frac{x}{y}$.

Exercise 5.17.3 Suppose $\frac{a}{b} \sim \frac{c}{d}$ and neither a nor c is 0. Show $\frac{b}{a} \sim \frac{d}{c}$.

Exercise 5.17.4 If $Y \neq \left[\frac{0}{1}\right]$, show $X \div Y = X \cdot Y^{-1}$.

Exercise 5.17.5 If $Y \neq \left[\frac{0}{1}\right]$, show $Y \cdot Y^{-1} = \left[\frac{1}{1}\right]$.

Exercise 5.17.6 If $Y \neq \left[\frac{0}{1}\right]$, show $\left(Y^{-1}\right)^{-1} = Y$.

Exercise 5.17.7 Show $(X \cdot Y)^{-1} = X^{-1} \cdot Y^{-1}$.

Exercise 5.17.8 Show $\left[\frac{a}{1}\right] \div \left[\frac{b}{1}\right] = \left[\frac{a}{b}\right]$.

5.18   Simplifying our notation

Fractions and rational numbers are different things. Rational numbers are the things we are really interested in; fractions are merely names for them. But fractions are important because it is by manipulating them that we work with the rational numbers they name. Our notation thus far has kept this distinction visible, allowing us to tell at a glance whether we are talking about a fraction, $\frac{1}{2}$ say, or a rational number, $\left[\frac{1}{2}\right]$ say. But now the basic properties of the rational number system have been developed, and we can safely stop pointing out the distinction.

From now on we will use the symbol $\frac{a}{b}$ for *both* the fraction $\frac{a}{b}$ and the rational number $\left[\frac{a}{b}\right]$. It can be told from context which is meant.

For example, if we write $\frac{1}{2} = \frac{2}{4}$ we are clearly talking about rational numbers, and in earlier sections we would have written $\left[\frac{1}{2}\right] = \left[\frac{2}{4}\right]$. But if we write that $\frac{1}{2}$ is in lowest terms, we must be talking about a fraction, since rational numbers, being sets of fractions, don't themselves have tops and bottoms to work with.

Exercises

Exercise 5.18.1 Which of the following statements are about fractions, which are about rational numbers?

1. $\frac{a}{b} = \frac{n \cdot a}{n \cdot b}$.

2. $\frac{3}{4} + \frac{3}{4} = \frac{3}{2}$.

3. $\frac{2}{3}$ has a denominator of 3.

4. 12 is a common denominator for $\frac{2}{3}$ and $\frac{3}{4}$.

5. The inverse of $\frac{2}{3}$ is $\frac{3}{2}$.

5.19 Whole numbers and rational numbers

No rational number is also a whole number; they are quite different things. To work with rational numbers, we use fractions, and to work with fractions, we must already know how to work with whole numbers. To say whole numbers are 'among' the rationals would lead to a circular situation. But there are certain rational numbers that 'behave like' whole numbers. In this section we take a close look at this.

To say a line segment is $\frac{3}{1}$ Unit just says it is 3 Units long. Having realized this, we might expect $\frac{3}{1}$ and 3 to have similar properties in their respective number systems. More generally, let us say the whole number a and the rational number $\frac{a}{1}$ *correspond*. We will see that whole numbers and their corresponding rational numbers behave alike in adding, multiplying, comparing, subtracting and dividing.

First we look at addition. Suppose, in the whole number system, $a + b = c$. To say the corresponding rationals behave the same way in the rational number system is to say $\frac{a}{1} + \frac{b}{1} = \frac{c}{1}$. But, in fact, $\frac{a}{1} + \frac{b}{1} = \frac{a \cdot 1 + 1 \cdot b}{1 \cdot 1} = \frac{a+b}{1} = \frac{c}{1}$. This may be succinctly expressed as $\frac{a}{1} + \frac{b}{1} = \frac{a+b}{1}$.

Exercise 5.19.1 says that a and $\frac{a}{1}$ behave alike in their respective number systems. Since we are finished with our development of the properties of rational numbers it will do no harm, from now on, to stop emphasizing the distinction between 3 and $\frac{3}{1}$ say.

From now on we will use the symbol a for *both* the whole number a and the rational number $\frac{a}{1}$. Either the context will make it clear which is meant, or else it won't matter.

For example, we may write $\frac{3}{2} + \frac{5}{2} = 4$ instead of $\frac{3}{2} + \frac{5}{2} = \frac{4}{1}$.

Notice that 0 and $\frac{0}{1}$ get identified. But, in fact, Exercise 5.8.3 says $\frac{0}{1}$ behaves like an 'additive identity' for rationals. Also 1 and $\frac{1}{1}$ get identified, and we have Exercise 5.11.5 which says $\frac{1}{1}$ behaves like a 'multiplicative identity' for rationals.

We call attention one last time to the fact that 4 and $\frac{4}{4}$ are different things. But once it has been proved that they have similar properties, it is harmless to write them both the same way.

Exercises

Exercise 5.19.1 Show:

1. $\frac{a}{1} \cdot \frac{b}{1} = \frac{a \cdot b}{1}$;

2. $a > b$ if and only if $\frac{a}{1} > \frac{b}{1}$;

3. $\frac{a}{1} - \frac{b}{1} = \frac{a-b}{1}$

4. if a is divisible by b, $\frac{a}{1} \div \frac{b}{1} = \frac{a \div b}{1}$.

5.20   A different notation for division

It is easy to see that $\frac{a}{1} + \frac{b}{1} = \frac{a}{b}$ or, using our abbreviated notation, $a \div b = \frac{a}{b}$. Thus, for *whole numbers* we, in effect, have two notations for division. It is reasonable to extend this dual notation to all rational numbers.

Definition 5.20.1 Let A and B be rational numbers with $B \neq 0$. We write $\frac{A}{B}$ for $A \div B$, that is, for $A \cdot B^{-1}$.

It might be asked, what was wrong with the notation $A \div B$ for division? The answer is that nothing is wrong with it. But it turns out that the bar indicating division of rational numbers and the bar of a fraction have similar formal properties. Then, we need not learn one set of rules for handling fractions and another for dividing; both sets of rules will look the same.

For example, by definition and Theorem 5.4.2, $\frac{x}{y} = \frac{z}{w}$ if and only if $x \cdot w = y \cdot z$. But also,

Theorem 5.20.2 $\frac{X}{Y} = \frac{Z}{W}$ *if and only if* $X \cdot W = Y \cdot Z$ *(where* X, Y, Z, *and* W *are any rational numbers, with* Y *and* W *not 0).*

Proof Suppose $\frac{X}{Y} = \frac{Z}{W}$ (which means $X \cdot Y^{-1} = Z \cdot W^{-1}$). Then $X \cdot Y^{-1} \cdot Y = Z \cdot W^{-1} \cdot Y$ or by Exercise 5.17.5, $X \cdot 1 = Z \cdot W^{-1} \cdot Y$ or, by Exercise 5.11.5, $X = Z \cdot W^{-1} \cdot Y$. Then, similarly, $X \cdot W = Z \cdot W^{-1} \cdot Y \cdot W = Z \cdot Y \cdot W^{-1} \cdot W = Z \cdot Y \cdot 1 = Z \cdot Y$. Thus $X \cdot W = Z \cdot Y$. □

Other similarities in the behavior of the bar of fractions and the bar for division are listed in the following theorems.

Theorem 5.20.3

1. $\frac{X}{Y} = \frac{X \cdot W}{Y \cdot W}$.

2. $\frac{X}{W} + \frac{Y}{W} = \frac{X+Y}{W}$.

3. $\frac{X}{Y} + \frac{Z}{W} = \frac{X \cdot W + Y \cdot Z}{Y \cdot W}$.

Theorem 5.20.4 $\frac{X}{Y} > \frac{Z}{W}$ *if and only if* $X \cdot W > Y \cdot Z$.

Theorem 5.20.5

1. $\frac{X}{Y} \cdot \frac{Z}{W} = \frac{X \cdot Z}{Y \cdot W}$.

2. $\frac{X}{Y} \div \frac{Z}{W} = \frac{X}{Y} \cdot \frac{W}{Z}$.

Exercises

Exercise 5.20.1 Complete the proof of Theorem 5.20.2 by showing: if $X \cdot W = Y \cdot Z$ then $\frac{X}{Y} = \frac{Z}{W}$.

Exercise 5.20.2 Prove Theorem 5.20.3.

Exercise 5.20.3 Prove Theorem 5.20.4.

Exercise 5.20.4 Prove Theorem 5.20.5.

CHAPTER 6

Finite Decimals

What is it that is here propounded? Some wonderful invention? Hardly that, but a thing so simple that it scarce deserves the name invention; for it is as if some stupid country lout chanced upon great treasure without using any skill in the finding...

–La Disme
Simon Stevin
(1548 - 1620)

6.1 Introduction

The system of rational numbers is essential; it is difficult to imagine our world trying to get along without it. But there are drawbacks in working with fractions. For instance, Archimedes once proved that π was between $3\frac{1335}{9347}$ and $3\frac{1337}{8069}$. Question: which of these is bigger? The trouble is, we have no tool available for the rational numbers comparable to place-value notation for the whole numbers.

A long time ago, people realized that things were simpler if they only worked with fractions in which the denominators were powers of some single, fixed number. For instance, it is customary to divide inches into halves, quarters, eighths, etc., so that all measurements are expressed using fractions having powers of 2 in the denominator.

At some point it began to be accepted that a good choice for this key number was 10, the base of our standard system of notation. Fractions whose denominators are powers of 10 are called *decimal fractions*. Finally, in 1585, Simon Stevin published a book, *La Disme*, in which he set out a convenient system of naming decimal fractions, essentially what we now call *finite decimals*.

In this chapter we will study a notion of *base n finite decimal* for each base. It turns out that in no base do we have more than a portion of the rational numbers available. Yet base n finite decimals provide an entirely satisfactory system for everyday use. We will see why. Also, on a more abstract level, the work done in this chapter provides the foundations for our development of the system of *real numbers* in later chapters.

6.2   Finite decimals

Here, as in Chapter Four, when we refer to *base n* we always assume n is a
whole number ≥ 2.

Definition 6.2.1 A *base n decimal fraction* is a fraction in which the denominator is a power of n; that is, it is a fraction of the form $\frac{a}{n^b}$ where a and b are
whole numbers.

Example (in which we use conventional base 10 names to name whole num-
bers). $\frac{1}{2}$, $\frac{1}{4}$, $\frac{3}{8}$, $\frac{7}{1}$ are all base 2 decimal fractions. $\frac{1}{10}$, $\frac{5}{10}$, $\frac{7834}{100}$, $\frac{7}{1}$ are all base
10 decimal fractions.

A few observations are in order. First, and least important, the terminology
is somewhat unfortunate. The word 'decimal' literally refers to *ten*, and it is
awkward to be using it when other bases are involved. We have to stretch
words a bit in order to discuss things that common language has not provided
for. Incidentally, this point came up earlier too. 'Digit' not only has numerical
significance, but (not coincidentally) it means *finger*. The canonical human
being has ten digits, yet we felt free to talk about base 12 digits.

On a more mathematical level, since $n^0 = 1$ for all bases n, then fractions
with denominators of 1 are base n decimal fractions for all n. These are the
ones that behave like whole numbers.

Finally, it is *fractions* that we are talking about. Thus $\frac{1}{2}$ is a base 2 decimal
fraction while the different fraction $\frac{5}{10}$ is a base 10 decimal fraction. As usual,
though, it is rational numbers we are really interested in. When we say some
rational number can be expressed as a base n decimal fraction, we mean that
one of the fractions that names it is a base n decimal fraction. Thus the rational
number $\frac{1}{2}$ can be expressed both as a base 2 and as a base 10 decimal fraction.

Now we introduce a simple system for *naming* those rational numbers that
can be expressed as base n decimal fractions. Then we will work with the
names, much as we did when place-value notation for whole numbers was introduced in Chapter Four.

Definition 6.2.2 A *base n finite decimal* is an ordered pair of base n names for
whole numbers (improper names are allowed). The customary way of writing
a base n finite decimal is $w.z$ where w and z are the two base n whole number
names. In the base n finite decimal $w.z$, w is called the *whole number part* and
z the *decimal part*.

Example 101.011 is a base n finite decimal for every n. The whole number
part is 101 and the decimal part is 011 (an improper name).

Next we say how base n finite decimals can be thought of as naming numbers.

Definition 6.2.3 Let $w.z$ be a base n finite decimal, with z a base n name of length c. By $(w.z)_n$ we mean the *rational number*

$$(w)_n + \frac{(z)_n}{n^c}.$$

Example This is a base 10 example. For reading ease we omit subscripts on whole number names. All whole number names in this example are used in the base 10 sense. Now 23.168 is a base 10 finite decimal, the decimal part, 168, is of length 3, so

$$\begin{aligned}(23.168)_{10} &= 23 + \frac{168}{10^3}\\ &= 23 + \frac{168}{1000}.\end{aligned}$$

Some conventions. If a finite decimal $w.z$ has a decimal part of length c, we will call it a *c-place decimal*. Thus 23.168 is a 3-place decimal.

In a base n finite decimal, if the decimal part is 0 it, and the decimal point, are often omitted. Thus 23 may be written for 23.0. Likewise, if the whole number part is 0, it (but not the decimal point) may be omitted. Thus .168 may be used for 0.168. We should also note that some people say the decimal part of 23.168 is .168. For us, it is more convenient and natural not to include the decimal point itself in the decimal part; but rather treat the decimal point as punctuation, or a separating device, comparable in its role to the bar of a fraction.

The definition gives $(w.z)_n$ as a sum of two base n decimal fractions, which can be combined into a single one according to Exercise 6.2.1. In fact, this can always be done in a simple way. If we continue the previous example, we find

$$\begin{aligned}(23.168)_{10} &= 23 + \frac{168}{1000}\\ &= \frac{23 \cdot 1000}{1000} + \frac{168}{1000}\\ &= \frac{23000}{1000} + \frac{168}{1000}\\ &= \frac{23168}{1000}.\end{aligned}$$

Note that the numerator is simply the whole number part followed directly by the decimal part. Indeed, this always happens.

Theorem 6.2.4 *Let $w.z$ be a base n finite decimal. Then $(w.z)_n = \frac{(wz)_n}{n^c}$ where c is the length of z.*

Remark $(w.z)_n$ is defined earlier in this section. wz, being the concatenation of two base n whole number names, is another such, and hence $(wz)_n$ was defined in Chapter Four.

Proof By induction on the length of z, the decimal part. We leave the initial step where the length is 1 to you, and go straight to the induction step. So, suppose the result is known for all base n finite decimals whose decimal part is of length c. And suppose now we have one with a decimal part of length $c + 1$ to work with. Then the decimal part must be of the form zd, where z is of length c and d is a base n digit. And then, using the definitions from this section and from Chapter Four,

$$
\begin{aligned}
(w.zd)_n &= (w)_n + \frac{(zd)_n}{n^{c+1}} \\
&= (w)_n + \frac{(z)_n \cdot n + d}{n^{c+1}} \\
&= (w)_n + \frac{(z)_n \cdot n}{n^{c+1}} + \frac{d}{n^{c+1}} \\
&= (w)_n + \frac{(z)_n}{n^c} + \frac{d}{n^{c+1}} \\
&= (w.z)_n + \frac{d}{n^{c+1}} \\
&= \frac{(wz)_n}{n^c} + \frac{d}{n^{c+1}} \qquad \text{(by induction hypothesis)} \\
&= \frac{(wz)_n \cdot n}{n^{c+1}} + \frac{d}{n^{c+1}} \\
&= \frac{(wz)_n \cdot n + d}{n^{c+1}} \\
&= \frac{(wzd)_n}{n^{c+1}}
\end{aligned}
$$

This concludes the induction step. □

Finally, we have said that base n finite decimals and base n decimal fractions were different ways of designating the same rational numbers. We sketch a proof of this now.

Theorem 6.2.5 *A rational number can be expressed as a base n decimal fraction if and only if it can be named by a base n finite decimal.*

Proof Suppose we have a base n finite decimal, say $w.z$. Then

$$(w.z)_n = (w)_n + \frac{(z)_n}{n^c}$$

and the right hand side is the sum of two (rationals named by) base n decimal fractions, and hence is itself expressible as a base n decimal fraction, by Exercise 6.2.1.

In the other direction, suppose we have the base n decimal fraction $\frac{x}{n^c}$. Since $\frac{x}{n^c} = \frac{x \cdot n}{n^{c+1}}$, we can always arrange things so that the exponent in the denominator is a counting number, and hence can be the length of a whole number name. We assume this is the case in what follows. Let y be a base n name for the whole number x, thus $(y)_n = x$. By using Theorem 4.2.6 we can always find such a y whose length is at least $c + 1$; let us say we have done so. Then y can be written as wz where z is a name of length c and w is a name of length at least 1. But then, using Theorem 6.2.4,

$$\frac{x}{n^c} = \frac{(wz)_n}{n^c} = (w.z)_n$$

hence we have a base n finite decimal name too. ☐

Exercises

Exercise 6.2.1

1. Show that the sum of two base n decimal fractions is equivalent to a base n decimal fraction. Similarly for the difference, when defined.

2. Show that the product of two base n decimal fractions is another base n decimal fraction.

Exercise 6.2.2 Show the result stated in Theorem 6.2.4 is correct when z is of length 1.

Exercise 6.2.3 Show $(w.z0)_n = (w.z)_n$.

6.3 Why finite decimals are not adequate

The base n finite decimals constitute a system of *names* for certain rational numbers. But we will be somewhat sloppy, and also call a *number* a base n finite decimal if it has a base n finite decimal name. It will be clear from context when we mean the number and when we mean the name.

Now the inadequacies of finite decimals are simply stated. No matter what base we choose, there are many rational numbers that aren't base n finite decimals. It follows from this that the base n finite decimals are not closed under division. These are pretty serious flaws.

Example In base 10, $\frac{1}{3}$ is not a finite decimal.

It is common knowledge that the 'usual' process for turning $\frac{1}{3}$ into a finite decimal produces an endless string of 3's. But since we have not discussed this technique, we have to follow a somewhat different, though equivalent approach. In the rest of this example *base 10 notation* is used throughout.

Suppose $\frac{1}{3}$ were some base 10 finite decimal, say $\frac{1}{3} = (w.z)_n$. Then by Theorem 6.2.4, $\frac{1}{3} = \frac{(wz)_n}{10^c}$ (where c is the length of z, hence c is at least 1). Now $(wz)_n$ is some whole number, let us write x for it. Thus

$$\frac{1}{3} = \frac{x}{10^c}$$

from which it follows that $3 \cdot x = 10^c$. This says 3 exactly divides 10^c. The following, together with Theorem 3.10.1, says this is not possible: division of 10^c by 3 always leaves a remainder of 1.

Fact For each counting number c there is a whole number q such that $3 \cdot q + 1 = 10^c$.

Proof By induction on c. If c is 1, simple calculation shows that taking $q = 3$ will do.

Suppose we have a whole number q such that $3 \cdot q + 1 = 10^c$. We produce a whole number q' such that $3 \cdot q' + 1 = 10^{c+1}$ which will complete the proof. Well, let $q' = q \cdot 10 + 3$. Then

$$
\begin{aligned}
3 \cdot q' + 1 &= 3 \cdot (q \cdot 10 + 3) + 1 \\
&= 3 \cdot q \cdot 10 + 3 \cdot 3 + 1 \\
&= 3 \cdot q \cdot 10 + 10 \\
&= (3 \cdot q + 1) \cdot 10 \\
&= 10^c \cdot 10 \\
&= 10^{c+1}
\end{aligned}
$$

□

One quick consequence of this is that the result of dividing two finite decimals need not be another finite decimal. (Division is meaningful, though, since finite decimals are just certain rational numbers, and division for them was defined in the previous chapter.) Consider the base 10 problem: $1.0 \div 3.0$. Now, $1.0 = 1 + 0/10^1 = 1$ and $3.0 = 3 + 0/10^1 = 3$ so $1.0 \div 3.0 = 1 \div 3 = 1/3$ and we just saw that $1/3$ is not any base 10 finite decimal.

A topic we have not discussed yet is *changing bases* when finite decimals are involved. This is quite simple though since both the definition of $(w.z)_n$ and Theorem 6.2.4 involve whole numbers, and we know how to handle them. For example, let us convert $(3.23)_5$ to base 10. Well, using Theorem 6.2.4 (and Exercise 4.2.4)

$$(3.23)_5 = \frac{(323)_5}{(100)_5}$$

Now by the conversion techniques for whole number names from Chapter Four we have $(323)_5 = (88)_{10}$ and $(100)_5 = (25)_{10}$ so

$$(3.23)_5 = \frac{(88)_{10}}{(25)_{10}}$$

$$= \frac{(88)_{10} \cdot (4)_{10}}{(25)_{10} \cdot (4)_{10}}$$

$$= \frac{(352)_{10}}{(100)_{10}}$$

$$= (3.52)_{10}.$$

But conversion of a finite decimal does not always produce a finite decimal. For example, suppose we try converting $(0.4)_{12}$ to base 10.

$$(0.4)_{12} = \frac{(4)_{12}}{(10)_{12}}$$

$$= \frac{(4)_{10}}{(12)_{10}}$$

$$= \frac{(1)_{10} \cdot (4)_{10}}{(3)_{10} \cdot (4)_{10}}$$

$$= \frac{(1)_{10}}{(3)_{10}}$$

And our earlier example showed that this could not be written as a base 10 finite decimal.

Finally we show that examples like $\frac{1}{3}$ in base 10 occur for every base choice. Thus no matter what n we pick, the base n finite decimals never give us the whole system of rational numbers to work with.

Theorem 6.3.1 *For each base, n, the rational number $\frac{1}{n+1}$ is not a base n finite decimal.*

Proof Suppose $\frac{1}{n+1}$ were a base n finite decimal. We derive a contradiction. If $\frac{1}{n+1}$ were a base n finite decimal then it could be written as a base n decimal fraction, say as

$$\frac{1}{n+1} = \frac{x}{n^c}$$

for some whole numbers x and c. Now c cannot be 0 since if it were, $\frac{x}{n^c} = x$ so $\frac{1}{n+1}$ would be a whole number. But since n is a base, $n > 1$ and it follows that

$$0 < \frac{1}{n+1} < \frac{1}{2} < 1$$

so $1 < 1 + \frac{1}{n+1} < 2$. If $\frac{1}{n+1}$ were a whole number, so would $1 + \frac{1}{n+1}$ be, and this would contradict Theorem 2.13.2.

Thus c must be a counting number. But then, many choices of c are possible, since

$$\frac{x}{n^c} = \frac{x \cdot n}{n^{c+1}}$$

Form a collection S of counting numbers as follows. Put c into S if, for some whole number x,

$$\frac{1}{n+1} = \frac{x}{n^c}.$$

Now suppose c is a member of S. Then for some x we have

$$\frac{1}{n+1} = \frac{x}{n^c}.$$

Then

$$
\begin{aligned}
n^c &= x \cdot (n+1) \\
&= x \cdot n + x \\
n^c - x \cdot n &= x \\
n \cdot (n^{c-1} - x) &= x
\end{aligned}
$$

If we write y for $n^{c-1} - x$ we have $n \cdot y = x$. Thus

$$
\begin{aligned}
\frac{1}{n+1} &= \frac{x}{n^c} \\
&= \frac{n \cdot y}{n^c} \\
&= \frac{y}{n^{c-1}}.
\end{aligned}
$$

As we showed earlier, for this to happen, $c - 1$ must be a counting number. But then the criteria is met for $c - 1$ to also be a member of S.

We have shown that S can have no least member. Then the Well Ordering Theorem, 2.13.1, says S must be empty. Hence $\frac{1}{n+1}$ cannot be written as a base n decimal fraction. $\square$

Exercises

Exercise 6.3.1

1. Show $\frac{1}{2}$ is not a *base 5* finite decimal.

2. Show $\frac{1}{5}$ is not a *base 12* finite decimal.

Exercise 6.3.2 Convert the following base 5 finite decimals to base 10 finite decimals.

1. 14.21;

2. 2.001;

3. 0.03.

Exercise 6.3.3 Convert the following base 12 finite decimals to base 10 fractions. Which of them can be written as base 10 finite decimals?

1. $(0.6)_{12}$;

2. $(0.8)_{12}$;

3. $(0.5)_{12}$.

Exercise 6.3.4 Prove that any given rational number $\frac{a}{b}$ is a finite decimal in *some* base.

6.4 Why finite decimals are adequate

> Tom...found himself swimming about in the stream, being about four inches, or – that I may be accurate – 3.87902 inches long...
>
> –The Water Babies,
> Charles Kingsley, 1863

Not every rational number is a finite decimal, no matter what base we use. But, as we will show, there are finite decimals as close as we need to any given rational number. This means finite decimals can have great everyday utility. After all, no measurement is perfect, it is always to the nearest thousandth of an inch, or the nearest millionth, or some such. But then if measurements are all approximate things, which can be expressed by finite decimals, and if rational numbers can be suitably approximated with finite decimals, then finite decimals are all that need enter into our everyday calculations. Finite decimals are enough, as the prevalence of pocket calculators indicates.

What we will show is that. for any rational number x, there is a base n finite decimal as close to x as we want. That is, if you specify a 'degree of closeness' (some small rational number ϵ *other than 0*) we can find a base n finite decimal within ϵ of x.

To begin with, we simplify our job a little. The degree of closeness, ϵ, can be any non-zero rational number, say $\frac{y}{z}$. We show it is enough to consider those in which the numerator is 1. Well, since ϵ, or $\frac{y}{z}$, is not 0, y is not 0. Then, using the Archimedean Order property of the whole number system, Theorem 3.8.2, there is a whole number m such that $m \cdot y > z$. From this we get that $\frac{1}{m} < \frac{y}{z}$. Consequently, if we can find a finite decimal within $\frac{1}{m}$ of x, we are also certain of being within $\frac{y}{z}$, or ϵ, of x too. Thus it is sufficient to consider only those 'degrees of closeness' that can be expressed in the form $\frac{1}{m}$, with a numerator of 1.

One more simplification before we get to the heart of the matter. Above we simplified the numerator of our 'degree of closeness' expression; now we simplify the denominator. We show it is enough to consider only those in which the denominator is a power of n.

Well, consider $\frac{1}{m}$. The whole number m has a base n name, say w. Thus $(w)_n = m$. Say the length of w is p. Now consider the base n name consisting of 1 followed by p 0's. This is a proper name of length $p + 1$ hence, by our methods of comparing numbers using place value names, it must name a bigger number than w names. But repeated use of Exercise 4.2.4 (actually an induction argument is involved) shows that 1 followed by p 0's is a base n name for n^p. Thus $n^p > (w)_n = m$. From this it follows that $\frac{1}{n^p} < \frac{1}{m}$.

Then, if we can find a finite decimal within $\frac{1}{n^p}$ of x, we are sure of being within $\frac{1}{m}$ of x as well. Thus we actually only consider those 'degrees of closeness' that can be expressed in the form $\frac{1}{n^p}$ for some p. Actually, this agrees with common practice where we talk about 'nearest hundredth', or 'nearest thousandth', etc.

Theorem 6.4.1 *Let $\frac{a}{b}$ be any rational number and let p be any whole number. There is a base n finite decimal f within $\frac{1}{n^p}$ of $\frac{a}{b}$. More precisely, there is a base n finite decimal f such that*

$$\frac{a}{b} - (f)_n < \frac{1}{n^p}.$$

Proof Long divide b into $a \cdot n^p$ (Theorem 3.10.1) getting quotient q and remainder r. Thus $a \cdot n^p = q \cdot b + r$ and $r < b$. Now

$$\frac{a}{b} = \frac{a \cdot n^p}{b} \cdot \frac{1}{n^p}$$

$$= \frac{q \cdot b + r}{b} \cdot \frac{1}{n^p}$$

$$= \left(\frac{q \cdot b}{b} + \frac{r}{b} \right) \cdot \frac{1}{n^p}$$

$$= \left(q + \frac{r}{b} \right) \cdot \frac{1}{n^p}$$

$$= q \cdot \frac{1}{n^p} + \frac{r}{b} \cdot \frac{1}{n^p}$$

$$= \frac{q}{n^p} + \frac{r}{b} \cdot \frac{1}{n^p}.$$

Now $\frac{q}{n^p}$ is a base n decimal fraction, and it differs from $\frac{a}{b}$ by $\frac{r}{b} \cdot \frac{1}{n^p}$. But recall, $r < b$ so

$$\frac{r}{b} < 1$$

$$\frac{r}{b} \cdot \frac{1}{n^p} < \frac{1}{n^p}.$$

So we have found a base n decimal fraction, and hence a base n finite decimal, within $\frac{1}{n^p}$ of $\frac{a}{b}$. This concludes the proof. $\square$

We should say something about how such approximations are calculated in practice. Consider the base 10 problem of finding a finite decimal within $\frac{1}{1000} = \frac{1}{10^3}$ of $\frac{1}{3}$. By the method of proof given above, we should long divide 3 into $1 \cdot 10^3 = 1000$. Using the customary arrangement, the work appears thus:

$$
\begin{array}{r}
333 \\
3)\overline{1000} \\
\underline{9} \\
10 \\
\underline{9} \\
10 \\
\underline{9} \\
1
\end{array}
$$

Note that this is a whole number problem. We have our quotient $q = 333$. Now the proof says the desired finite decimal is one corresponding to $\frac{333}{10^3}$ or, simply, .333.

In practice, the two stages (determining q and locating the decimal point) are combined into a single algorithm, and what one sees on paper is just

$$
\begin{array}{r}
.333 \\
3)\overline{1.000} \\
\underline{9} \\
10 \\
\underline{9} \\
10 \\
\underline{9} \\
1
\end{array}
$$

You should have no trouble in accounting for what is happening here. The approximation, .333, is generally referred to as a *3-place approximation*, for obvious reasons.

One last observation. We saw in the previous section that the result of dividing one base n finite decimal by another need not produce a base n finite decimal. It will, however, always produce a rational number, and this rational number can be approximated by base n finite decimals as closely as we like, using the method of this section. In fact, as the reader will readily see in the exercises, this is nothing more than what is usually called finite decimal division.

Exercises

Exercise 6.4.1

1. Find a 4-place approximation to $\frac{1}{11}$ in base 2.

2. Find a 3-place approximation to $\frac{1}{5}$ in base 12.

Exercise 6.4.2 Show that between two distinct rational numbers there is a base n finite decimal. Hint: By Exercise 5.13.5, between any two rational numbers a and b there is another *rational number c*. Now, approximate to c 'suitably closely.'

Exercise 6.4.3 In base 10, compute a 3-place approximation to the rational number $2.81 \div 3.656$.

Exercise 6.4.4 In base 2, compute a 5-place approximation to $11.01 \div 1.101$.

6.5   Basic arithmetic

> ...this discovery of decimal numbers...does away with all these difficulties. To speak briefly ...all computations of the type of the four principles of arithmetic – addition, subtraction, multiplication and division – may be performed by whole numbers with as much ease as in counter-reckoning.
>
> –La Disme
> Simon Stevin, 1585

As might be expected, all the elementary arithmetic algorithms are much simplified if finite decimals are used. We briefly sketch why.

Comparing numbers

Exercise 6.2.3 says we can always append 0's to the end of the decimal part of a finite decimal without affecting what it names. Consequently, we can always arrange things so that any two finite decimals we want to compare have the same length decimal parts. Also recall that in Chapter Four rules were presented for comparing *whole* numbers, using their base n names.

Now we give three different methods for comparing finite decimals. All are minor variations on the same theme, but the third, because of the way it is stated, can be generalized in a manner the other two can not. It will, in fact, form the basis of our treatment of real numbers in the next chapter.

Suppose $w_1.z_1$ and $w_2.z_2$ are two base n finite decimals, *with z_1 and z_2 of the same length.*

Method I If $(w_1 z_1)_n < (w_2 z_2)_n$ then $(w_1.z_1)_n < (w_2.z_2)_n$.

Example $(101.001)_2 < (101.010)_2$ because, as whole numbers, $(101001)_2 < (101010)_2$.

The justification for this method of comparison is simple. We use Theorem 6.2.4, and our methods of comparing fractions from Chapter Five. Say both z_1 and z_2 are of length k. Then $(w_1.z_1)_n < (w_2.z_2)_n$ is equivalent to $\frac{(w_1 z_1)_n}{n^k} < \frac{(w_2 z_2)_n}{n^k}$ which is equivalent to $(w_1 z_1)_n < (w_2 z_2)_n$.

Method II This time we need two rules.

1. If $(w_1)_n < (w_2)_n$ then $(w_1.z_1)_n < (w_2.z_2)_n$.

2. If $(w_1)_n = (w_2)_n$ but $(z_1)_n < (z_2)_n$ then $(w_1.z_1)_n < (w_2.z_2)_n$.

In words, rank finite decimals by their whole number parts if they are different, otherwise go by their decimal parts.

Method III Again there are two rules.

1. If $(w_1)_n < (w_2)_n$ then $(w_1.z_1)_n < (w_2.z_2)_n$.

2. If $(w_1)_n = (w_2)_n$ then, start at the decimal point and move digit by digit to the right to locate the first term where the two decimal parts have different digits. The one with the bigger digit there is the one that names the bigger number.

This method is just Method II again, but with clause 2 expanded by incorporating certain whole number comparison techniques.

Addition and Subtraction

Suppose we have two base n finite decimals that we want to add (more precisely, we want to find a name for the sum of the numbers they name).

To begin with, we can always arrange things so both finite decimals have the same length decimal parts, by appending 0's to the end of the shorter one if necessary. Let us suppose this has been done. Then, say we have the base n finite decimals $w_1.z_1$ and $w_2.z_2$, both k-place. Well, very simply, $(w_1.z_1)_n + (w_2.z_2)_n = (w_3.z_3)_n$ where $w_3.z_3$ is also a k-place finite decimal in which the string of digits, $w_3 z_3$, is the result of carrying out the whole number addition that arises if we simply ignore decimal points, that is, $(w_3 z_3)_n = (w_1 z_1)_n + (w_2 z_2)_n$.

The justification here is trivial. Using Theorem 6.2.4, and Theorem 5.6.1,

$$(w_1.z_1)_n + (w_2.z_2)_n = \frac{(w_1 z_1)_n}{n^k} + \frac{(w_2 z_2)_n}{n^k}$$
$$= \frac{(w_1 z_1)_n + (w_2 z_2)_n}{n^k}$$

and this is a base n decimal fraction that corresponds to the base n finite decimal $w_3.z_3$ described above.

As just described, there is a certain preparation step of appending 0's so that the two finite decimals have the same length decimal parts. In practice this is usually simplified to: "line up the decimal points." Then one treats the absence of a decimal part digit as the equivalent of a 0. Clearly this amounts to little more than a minor variation on what was described above.

Subtraction is handled in virtually the same way as addition (except that a whole number subtraction is performed, instead of a whole number addition). We skip details.

Multiplication

This time it is *not* desirable to arrange things so that decimal parts are the same length. Now, the rule is as follows.

$$(w_1.z_1)_n \cdot (w_2.z_2)_n = (w_3.z_3)_n$$

where

1. $w_3 z_3$ arises from a whole number calculation in which decimal points are ignored, that is, $(w_3 z_3)_n = (w_1 z_1)_n \cdot (w_2 z_2)_n$, and

2. the length of z_3 is the length of z_1 plus the length of z_2.

The justification is quite straightforward, using Theorem 2.10.2. Say z_1 is of length k and z_2 is of length p. Then

$$(w_1.z_1)_n \cdot (w_2.z_2)_n = \frac{(w_1 z_1)_n}{n^k} \cdot \frac{(w_2 z_2)_n}{n^p}$$
$$= \frac{(w_1 z_1)_n \cdot (w_2 z_2)_n}{n^k \cdot n^p}$$
$$= \frac{(w_3 z_3)_n}{n^{k+p}}$$

Exercises

Exercise 6.5.1 Justify Method II.

Exercise 6.5.2 Add, in base 5,

1. $14.312 + 1.13$

2. $.013 + .102$

Exercise 6.5.3 Add, in base 2,

1. $10.101 + 1.1$

2. $10.01 + .101$

Exercise 6.5.4 Add, in base 12,

1. $9.te + 1.9$

2. $10.1t + .03$

Exercise 6.5.5

1. Subtract, in base 5, $14.2\jmath3 - 3.41$

2. Subtract, in base 2, $101.11001 - 11.111011$

Exercise 6.5.6

1. In base 5, multiply $.134 \times 2.01$

2. In base 2, multiply 1.01×11.001

3. In base 12, multiply $1.t2 \times te.001$

CHAPTER 7

Real Numbers

7.1 Introduction

In finite decimals we have an ideal number system for everyday. It is easy to use, and we can be as accurate as we need to be. But this is not a good system for the general development of mathematics; there just aren't enough numbers.

We illustrate this with an ancient example. Suppose we have a square one foot on a side; how long is the diagonal? Now any 'square' we could actually draw would only be an approximate square since our instruments and tools for drawing, however good, are not perfect. And we can only measure the diagonal of a 'square' we have drawn to within the limits of accuracy of our measuring equipment, to the nearest millionth of an inch, say, and finite decimals will suffice for expressing the result of such a measurement. But in mathematics we do not work this way. Rather we imagine ideal, perfect squares, and we deduce our results about them from basic principles of geometry. We may draw pictures to help us think but we don't reach our results by measuring them. Now, in this ideal sense, how long is the diagonal of a square one foot on a side?

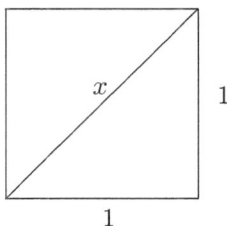

By the Pythagorean theorem, $x^2 = 1^2 + 1^2$, so $x^2 = 2$. That is, the length of the diagonal is a number whose square is 2. But, *there is no such number in any number system we have studied thus far*. Since all the kinds of numbers we have seen thus far have their counterparts in the system of rational numbers, what we are really saying is: there is no rational number whose square is 2. We give a proof of this that is 2500 years old.

First we need a precise definition of odd numbers.

Definition 7.1.1 An odd number is one more than an even number. An even number is twice some whole number.

Then $2n$ is always an even number, for any whole number n, and $2n$ can represent any even number we like by an appropriate choice of n. Then too, $2n + 1$ is always an odd number, and any odd number can be put in this form by properly choosing a whole number n.

Lemma 7.1.2 *The square of an odd number is odd.*

Proof Pick any odd number. It can be represented as $2n + 1$. Square it $(2n + 1)^2 = 4n^2 + 4n + 1 = 2(2n^2 + 2n) + 1$. Now $2(2n^2 + 2n)$ is even, since it is twice a whole number, so $2(2n^2 + 2n) + 1$ is odd since it is one more than an even number. □

Theorem 7.1.3 *There is no rational number whose square is 2.*

Proof Suppose there did exist a rational number whose square was 2. We derive a contradiction from this, which shows it is impossible.

If there were a rational number whose square was 2, it could be named by a fraction. Form a collection, S, of whole numbers as follows. Put a whole number y into S if y is the denominator of some fraction x/y whose square is (equivalent to) 2. By our supposition, S is not empty, so by the Well-Ordering Theorem, S has a *least* member, call it b. Then, there is a fraction a/b whose square is 2, but there is no fraction with a smaller denominator whose square is 2. In particular, then, not both a and b can be even numbers, because if they were, we could 'cancel' a factor of 2 in the numerator and denominator of a/b to get an equivalent fraction with a smaller denominator, whose square would also be 2.

To sum up things thus far: we have a fraction a/b whose square is 2, and not both a and b are even whole numbers.

We proceed from here as follows.

$$\left(\frac{a}{b}\right)^2 = \frac{a^2}{b^2}$$

so

$$\frac{a^2}{b^2} = 2$$

$$a^2 = 2b^2.$$

But $2b^2$ is even (it is twice something). So a^2 is even. Then a is even, for by the lemma, if a were odd, a^2 would also be odd. Since a is even, it is twice some whole number, say $a = 2k$.

Then $a^2 = (2k)^2 = 4k^2$. Substituting, $4k^2 = 2b^2$ and cancelling, $2k^2 = b^2$. But $2k^2$ is even, so b^2 is even. Then by the lemma again, b is also even. We have our contradiction: not both a and b are even, a is even, b is even. This is impossible, but it must happen if a/b existed, so it doesn't. This concludes the proof. $\square$

It turns out this is not an isolated situation. If rational numbers are the only numbers we have, there are many geometric line segments we can't assign a numerical length to. In fact, in a sense that can be made precise in *set theory*, there are more lines we can't measure with rationals than there are lines we can.

Greek mathematicians 2500 years found it a disturbing result that there is no rational number whose square is 2. Originally, Greek mathematics was based on the notion that rational numbers were enough to measure all line segments. When the proof above was discovered it led to a crisis in Greek mathematics which was finally straightened out when Eudoxus (c.408 - c.355 BC) created his theory of proportions, the details of which we won't go into here.

Basically there are only two ways out of the difficulty; create a larger number system, or accept that there are not enough numbers and develop mathematics without them. We do the first, the Greeks did the second. As a result, High School geometry and Greek geometry are not really the same. Take the Pythagorean theorem as an example: the square of the hypotenuse of a right triangle is equal to the sum of the squares of the other two sides. We take this to mean: $c^2 = a^2 + b^2$ where c, a and b are *numbers*, measuring the lengths of the hypotenuse and the sides. A Greek would understand this to say: squares drawn on the sides of a right triangle can be combined into a square equal to one drawn on the hypotenuse. The difference here is noticeable but not great. But for more recent mathematics the two approaches would lead to vastly different formulations. Indeed, probably much present day mathematics would never have been created if the Greek approach were still followed.

It is our job in this chapter to present the modern way out of the difficulty; to create a larger number system. The system is called the *real number system*. The name is a little misleading, since they are no more real than other kinds of numbers. The name dates from a time when complex numbers were not well understood, and there was thought to be a great gulf between 'real' results and 'imaginary' ones.

In the 19th century real numbers were first defined properly and studied rigorously, though they had been used informally for centuries before that time. Two equivalent approaches, one due to Dedekind (1831-1916) and one due to Cantor (1845-1918) were developed. We have chosen to follow yet a third approach, defining real numbers to be infinite decimals. We feel this is an easier approach to follow, but unlike the others, it doesn't go directly from rational numbers to real numbers, but rather from rationals to finite decimals to real numbers. This explains the presence of Chapter 6. Properly seen,

though, our approach is really a variant of Cantor's, but we do not develop the
relationship here.

Exercises

Exercise 7.1.1 Using the definition, prove that every whole number is either
odd or even, and no whole number can be both.

Exercise 7.1.2 Above we proved $\sqrt{2}$ was not a rational number. Now prove
$\sqrt{3}$ is not a rational number in the following analogous way. First call a number
a *triple* (analog of even) if it is of the form $3n$. Call a number a *non-triple*
(analog of odd) if it isn't a triple, and hence if it is one of the forms $3n + 1$ or
$3n + 2$.

1. Prove: the square of a non-triple is a non-triple.

2. Now, if there were a rational whose square was 3, it could be represented
   as a fraction in lowest terms. Suppose there is a fraction a/b such that

$$\frac{a^2}{b^2} = 3$$

   and a and b have no common factors (in particular, not both a and b are
   triples). Derive a contradiction.

Exercise 7.1.3 Prove $\sqrt{5}$ is not a rational number.

Remark The following general result (or rather, its geometric counterpart)
was known to Euclid: if $\sqrt{n}$ is not an integer, it is not a rational number
either.

7.2   A word of warning

The uses of real numbers are quite different from the uses of finite decimals.
Finite decimals simplify calculations, but real numbers simplify theories. For
example, with real numbers available we can say every line segment has a
numerical length, and then many geometric properties become simple counterparts of elementary number facts. For this we need the existence of the entire
real number system and its general properties.

But real numbers are *not* everyday tools. We do not use them for calculation
the way we use rational numbers. In fact, if real numbers turn up at all we
generally replace them with rational numbers that are 'sufficiently close.' We
use finite decimal approximations to infinite decimals when we build bridges.

To give a familiar example, the entries found in trigonometry tables or given
by calculators are finite decimals. These are not the exact values, but they are
near enough for practical purposes. The true values are infinite decimals which

can not be written down. But the supposed existence of these infinite decimal values simplifies the theory of trigonometry considerably.

So we are no longer interested in computations, but in completeness for mathematical purposes. In fact, we will sometimes find ourselves beyond all possibility of computation ever by the kind of approximate methods we have been discussing. Some of our proofs below will conclude that a real number having certain properties exists, but they will provide no means of actually computing such a number or even approximating it. In some cases this won't be a defect of the particular proof; means of computing may not exist. This is inherent in the nature of the real number system. If we want a 'complete' number system with a reasonably uncomplicated theoretical structure, loss of computational ability is the price we must pay.

There are some mathematicians who object to such proceedings for philosophical reasons. They say it is meaningless to assert that a number exists unless there are means of computing it, or at least of computing a sequence of better and better approximations to it. To such mathematicians the general theory of real numbers we present is wrong.

Until recently, alternative theories of real numbers which are more constructive were very complicated and 'unnatural'. There is now a constructive alternate which works smoothly (see *Foundations of Constructive Analysis*, Errett Bishop, McGraw Hill, 1967, especially the Preface and Chapter 1). Time will tell to what extent it gains general acceptance. The theory we present is the one accepted and used by most mathematicians today.

7.3 Infinite decimals informally

We want to create an extension of the system of rational numbers in which, among other things, all numbers have square roots, and cube roots, and so on. To do this we must say what sort of things we are going to call numbers, and what we mean by addition, subtraction, multiplication, etc. The system we create is called the system of *real numbers*. We are going to take real numbers to be *infinite* decimals, and we will see that the choice of base is not important. In fact, we will develop systems of infinite decimals in all bases at the same time, but independently of each other. We begin our discussion with base 10, which is most familiar.

In the system of base 10 finite decimals, no number exactly represents $\frac{1}{3}$. But we can find a sequence of closer and closer approximations:

.3
.33
.333
.3333
etc.

Now we create a *new* object, an infinite decimal,

.3333...

which we can intuitively think of as a 'limit' to which the successive finite
decimal approximations above tend. (The string of dots is meant to indicate
that the sequence of digits goes on forever.) .3333... is *not* anything we have
discussed before. It is not a fraction, it is not a finite decimal. Still, it is a
rather natural sort of object to consider, and we have a reasonable idea of how
we want it to behave. For instance, we would want to define addition of infinite
decimals so that

$$.333\ldots + .333\ldots = .666\ldots$$

We will begin formal considerations of infinite decimals in the next section, for
now we carry on an informal discussion.

   If we assume we can operate with infinite decimals in a reasonable way,
seemingly strange things happen. For instance, we can show

$$.70000\ldots = .69999\ldots$$

To see this, suppose we denote .69999... by N:

$$N = .69999\ldots$$

One would expect that multiplying by 10 should move the decimal point one
place to the right. Thus

$$10N = 6.9999\ldots$$

$$100N = 69.999\ldots$$

Now, if we subtract $10N$ from $100N$, the decimal parts of 69.999... and
6.999..., being the same, can be expected to cancel. Thus

$$
\begin{array}{rcl}
100N & = & 69.000\ldots \\
-10N & = & -6.999\ldots \\
\hline
90N & = & 63.000\ldots
\end{array}
$$

so

$$
\begin{array}{rcl}
9N & = & 6.3000\ldots \\
N & = & .7000\ldots
\end{array}
$$

   A little more thought shows that this is not as strange as it seems to be
at first sight. Things must be possible with infinite decimals which have no
counterpart with finite decimals. For instance, if we wish to subtract the finite
decimal .001 from .700,

$$
\begin{array}{r}
.700 \\
- \quad .001 \\
\hline
\end{array}
$$

We might begin by borrowing, that is, we would rewrite .700 as .700 = .6(10)0 = .69(10). Then the problem becomes

$$
\begin{array}{r}
.69(10) \\
- \quad .001 \\
\hline
.699
\end{array}
$$

.69(10) is not a finite decimal in a strict sense, but only a temporary expression for the purpose of subtraction. But .7 = .70 = .700 = .7000 etc., and by borrowing as above we see that the following expressions all represent the same number:

.7
.6(10)
.69(10)
.699(10)
.6999(10)
etc.

We can think of these as tending to an infinite decimal, and clearly it has to be .69999....

Looking at it slightly differently, with the finite decimal .700 the 'most' borrowing we can do is .69(10), but with the infinite decimal .7000... we are permitted unlimited borrowing:

$$
\begin{array}{rcl}
.70000 & = & .6(10)000\ldots \\
& = & .69(10)000\ldots \\
& = & .699(10)000\ldots \\
& = & .6999(10)000\ldots
\end{array}
$$

etc.

If we imagine ourselves permitted infinitely many steps, we can rearrange .7000... into .6999....

Something else you might think about, if .69999... and .70000... are not to be equal, how far apart are they? This is related to the borrowing discussion we just presented.

We have not *proved* .7000... and .6999... are equal. Indeed we can not do so since we don't yet have any formal theory of infinite decimals. Still, all the things we did are things we would like to be able to do when such a theory is developed. We are forced to conclude that: *If we are to have a reasonable theory of (base 10) infinite decimals we must identify .69999... and .70000....*

This is not the only such situation. .360000... and .359999... will have to be identified, also 1.0000... and .999.... In fact, whenever an infinite decimal 'corresponds' to a finite decimal this will come up. And something similar happens in all other bases as well. Later, when we begin our formal development, we will have to deal with this problem at the start.

Exercises

Exercise 7.3.1 Give an informal argument that .36000... and .359999... must be identified (in base 10) by setting $N = .359999...$ and proceeding as we did above.

Exercise 7.3.2 In base 2, give an informal argument that 1.000... and .111... must be identified by setting $N = .111...$ and proceeding as we did above.

Exercise 7.3.3 In base 4 find another infinite decimal which ought to represent .210000....

Exercise 7.3.4 Find a base 5 fraction which ought to be equivalent to the base 5 infinite decimal .2444....

7.4   Infinite sequences

Starting with this chapter we make heavy use of what is essentially a new mathematical concept in this book, an *infinite sequence*. An infinite decimal will involve, in its decimal part, an infinite sequence of digits, and we will even need infinite sequences of infinite decimals. In this section we say briefly what infinite sequences are. We do not formally develop their properties; an informal description is enough for our purposes.

An *infinite sequence* is simply an arrangement of certain things in a first, second, third, etc. order. More precisely, it is an assignment of some object to each counting number. The thing assigned to 1 is called first, the thing assigned to 2 is called second, and so on. More precisely yet, an infinite sequence is simply a function whose domain is the entire collection of counting numbers (see Chapter 2.5).

We often use the following notation for infinite sequences. We denote the sequence itself by some letter, s say, and the first term of it (the object assigned by it to the number 1) by s_1, the second by s_2, etc. Schematically, we may write

$$s \text{ is } s_1, s_2, s_3, \ldots$$

If each of the terms of the infinite sequence is a member of a collection S, we simply say we have an infinite sequence of members of S.

A familiar example of an infinite sequence is provided by an infinite decimal. The decimal part may be considered to be an infinite sequence of digits. This is an example that will concern us for much of the rest of this book. If we are talking about such an infinite sequence we will generally omit commas in our schematic representation, writing just

$$s_1 s_2 s_3 \ldots$$

for the decimal part of an infinite decimal, taking s_1 to be the digit in the first decimal place, s_2 to be the digit in the second decimal place, and so on.

The Well Ordering Principle (Theorem 2.13.1) is used constantly in working with infinite sequences. Its use is so frequent that it is often overlooked. A typical example: suppose

$$s_1, s_2, s_3, \ldots$$

is an infinite sequence of counting numbers. If any of the terms are even, there must be a *first* even term of the sequence. The formal reasoning involved here is this: let $\mathbf{C}$ be the collection of counting numbers k for which s_k is even; we are supposing $\mathbf{C}$ is not empty, so $\mathbf{C}$ has a least member, say q. Then s_q is the first even term of the infinite sequence. It is this kind of reasoning that justifies statements like: if not every decimal place of an infinite decimal is 0, then there is a first non-zero decimal place.

A *finite sequence* is an arrangement of things in a first, second, ..., last order. More precisely, it is an assignment of things to some initial string of counting numbers, say to 1,2,3,...,n. We use similar notational conventions for finite sequences, for example writing $s_1, s_2, \ldots, s_n$ to denote the finite sequence whose first term is $s_1, \ldots$, and whose last term is s_n. We say the *length* of the finite sequence $s_1, s_2, \ldots, s_n$ is n.

If an infinite sequence is given, its first n terms constitute a finite sequence. If we are given the infinite sequence

$$s_1, s_2, s_3, \ldots$$

and we write

$$s_1, s_2, \ldots, s_n$$

we mean the finite sequence that resuls when only the first n terms of the infinite sequence are retained.

We use the same notational convention with finite sequences of *digits* that we do with infinite ones: we write them without commas thus: $s_1 s_2 \ldots s_n$. We assume throughout this chapter that finite sequences of digits, as just described, obey all the conditions for *whole number names* set forth in Chapter Four, using the 'natural' notion of concatenation. In a formal development of the sequence concept, this would need proof; we are taking it for granted in order that we may get on with the development of the real number system itself.

7.5 Infinite decimals formally

By a *base n infinite decimal* we mean an ordered pair consisting of a base n whole number name, and an infinite sequence of base n digits. We follow a notational convention similar to the one we used for finite decimals. Thus we write

$$w.a_1 a_2 a_3 \ldots$$

to denote the infinite decimal with whole number name w, and infinite sequence of digits $a_1 a_2 a_3 \ldots$. We call w the *whole number part* and $a_1 a_2 a_3 \ldots$

the *decimal part*. We speak of the first decimal place, the second decimal place, and so on, in the usual way.

For example, 3.1415926535... is a base 10 infinite decimal, and likewise 1101.111001011011... is a base 2 infinite decimal. But be careful; a base 2 infinite decimal is also a base 10 infinite decimal, but it means different things in the two bases, just as 11 names different whole numbers in base 2 and base 10. The choice of base must be clear each time. Either we will say in words what base we are using, or we will write something like $(1101.111001011011...)_2$ to indicate the base. If we just say 'infinite decimal' with no base mentioned, it means the exact base doesn't matter.

Please note that we are using the base-indicating subscripts somewhat differently than in earlier chapters. For finite decimals, $(10.01)_2$ *was* a certain rational number, the one named by the base 2 finite decimal 10.01. But for infinite decimals, the subscript in $(10.011...)_2$ merely tells us we will be using the decimal according to base 2 rules (yet to be specified).

Next we must take care of the problem raised earlier, that certain infinite decimals will have to be identified, for example, in base 10, .7000... and .6999.... We do this by choosing one as standard and thinking of the other as a non-standard version of it.

If a base 10 infinite decimal ends with an infinite sequence of 9's (for example, .69999...) we say it is in *non-standard form*. If we take an infinite decimal in non-standard form, change the digit just before the string of 9's to the next higher digit, and change all the 9's to 0's, we have *an infinite decimal in standard form*. (For example, putting .69999... into standard form gives us .70000...). If d is an infinite decimal in non-standard form and d' is the result of putting d into standard form, we say d' is *the standard form of d* and d *is the non-standard form of d'*. (For example, .7000... is the standard form of .69999..., and .69999... is the non-standard form of .70000...). If d' is the standard form of d, we write $d = d'$, taking them to be equal by definition. (For example, .69999... = .70000....)

We also introduce similar notions for other bases. For example, in base 5, 3.124444... is in non-standard form, while 3.13000... is the standard form of it.

Exercises

Exercise 7.5.1

1. Give an example of a base 2 infinite decimal in non-standard form.

2. Do the same for base 12.

Exercise 7.5.2 Give a proper definition of standard form for base n.

7.6 Finite decimals and infinite decimals

In Chapter 6 we discussed finite decimals, which were rational numbers of a particular kind. Now we are discussing infinite decimals, loosely infinite sequences of digits with a decimal point someplace. *No finite decimal is an infinite decimal.* They are different sorts of objects entirely. But, some infinite decimals will behave like finite decimals. For example, the finite decimal 3.279 and the infinite decimal 3.2790000... will behave similarly for computation purposes.

To keep the distinction between finite decimals and infinite decimals clear, we will continue to call expressions like 3.279 finite decimals, and we will introduce another name for expressions like 3.2790000....

Definition 7.6.1 A *base n terminating decimal* is a base n infinite decimal ending with an infinite sequence of 0's.

For example, 3.2790000... is a terminating decimal. Notice that terminating decimals are always in standard form. They are the only infinite decimals that also have non-standard forms.

We call 3.2790000... the terminating decimal *associated with* the finite decimal, 3.279. More generally,

Definition 7.6.2 Let d be some base n finite decimal. The terminating decimal produced by following d with an infinite string of 0's is called *the terminating decimal associated with d*, and is denoted by d^*.

For example, if $d = 3.279$ then $d^* = 3.2790000...$. Notice that 3.279, 3.2790, 3.27900, etc. all have the same terminating decimal associated with them.

One of the things we will have to prove is that finite decimals and their associated terminating decimals do, in fact, behave alike in their respective number systems.

7.7 Order of infinite decimals

In Section 6.5, we presented a discussion of the ordering of finite decimals. We are going to base our definition of an ordering relation for infinite decimals on that discussion, more precisely we generalize what we called Method III.

Let A and B be two distinct base n infinite decimals *in standard form*. We describe a procedure for comparing them.

First, if the whole number parts of A and B are different, we define the one with the bigger whole number part to be the bigger infinite decimal.

Second, if the whole number parts of A and B are the same, then starting at the decimal points of A and B, find the first decimal place where A and B have different digits. We define the one with the bigger digit to be the bigger infinite decimal.

If A is bigger than B, we write $A > B$, or $B < A$.

For example, in base 10, $31.6835\ldots > 12.9635\ldots$ because $31 > 12$. Also, in base 10, $7.63943\ldots > 7.62534\ldots$ because the whole number parts are the same, as are the first decimal places, but $3 > 2$ in the second decimal place. Likewise, in base 2, $101.1101101\ldots > 101.1011010\ldots$.

To compare infinite decimals not in standard form, first put them into standard form, then compare them.

Now we derive two basic properties of this ordering relation.

Theorem 7.7.1 (Trichotomy law) *For any base n infinite decimals A and B, exactly one of the following holds: $A > B$, $A = B$, $B > A$.*

Proof We may assume that A and B have already been put into standard form. Since they are base n infinite decimals they look like this:

$$A = w_A.a_1a_2a_3a_4\ldots$$

$$B = w_B.b_1b_2b_3b_4\ldots$$

where w_A and w_B are base n whole number names, and each a_i and b_i is a base n digit.

Suppose $A \neq B$; we show one of $A > B$ or $B > A$ must be the case. Now, if A and B are different, either they have different whole number parts, or else they differ at some decimal place. If w_A and w_B are different, either $w_A > w_B$ or $w_B > w_A$. In the first case $A > B$ and in the second $B > A$. Otherwise $w_A = w_B$. Then consider the decimal parts of A and B. If there is some decimal place at which A and B differ, there must be a first such place. Say A and B agree at each decimal place up to the ith and differ there; $a_i \neq b_i$. Either $a_i > b_i$ or $b_i > a_i$. In the first case, $A > B$, and in the second $B > A$.

So far we have shown one of $A > B$, $A = B$ and $B > A$ must happen. Now we show two of them can't happen at the same time.

Suppose $A > B$ and also $B > A$. If A and B have different whole number parts, since $A > B$ we would have $w_A > w_B$, and since $B > A$ we would have $w_B > w_A$. But this contradicts the trichotomy law for whole numbers. So the whole number parts must be the same. Yet $A > B$, so there must be a first decimal place where they are different, say the ith, $a_i \neq b_i$. Since $A > B$ we must have $a_i > b_i$, but also since $B > A$ we must have $b_i > a_i$, and this again contradicts the trichotomy law for whole numbers. Conclusion: we can't have $A > B$ and also $B > A$. $\square$

Theorem 7.7.2 (Transitivity) *Let A, B and C be base n infinite decimals. If $A > B$ and $B > C$ then $A > C$.*

Proof Again we may assume A, B and C have been put into standard form. Let us say

$$A = w_A.a_1a_2a_3a_4\ldots$$

$$B = w_B.b_1b_2b_3b_4\ldots$$

$$C = w_C.c_1c_2c_3c_4\ldots$$

where w_A, w_B and w_C are base n whole number names, and a_i, b_i and c_i are base n digits.

Suppose first that $w_A \neq w_B$. Since $A > B$ we must have $w_A > w_B$. Since also since $B > C$, w_C can't be smaller than w_B, so $w_B \geq w_C$. Then by transitivity for whole numbers, $w_A > w_C$, so $A > C$. A similar argument works if $w_B \neq w_C$.

Now suppose $w_A = w_B = w_C$. Since $A > B$, there is a first decimal place at which A and B differ, say the jth. Then $a_j > b_j$, but up to the jth place A and B are the same. Similarly since $B > C$ there is a first decimal place at which B and C differ, say the kth. Then $b_k > c_k$, but up to the kth place B and C are the same. Now by the trichotomy law for counting numbers we have three possibilities: $j = k$, $k < j$ or $j < k$.

Suppose first that $j = k$. Then A, B and C look like this:

$$A = w_A.a_1a_2\ldots a_{j-1}a_j\ldots$$

$$B = w_A.a_1a_2\ldots a_{j-1}b_j\ldots$$

$$C = w_A.a_1a_2\ldots a_{j-1}c_j\ldots$$

(do you see why?) Then A and C agree up to the jth place, but $a_j > c_j$ (since $a_j > b_j > c_j$, and we have transitivity for whole numbers). Then by definition, $A > C$.

Suppose next that $j < k$. Then A, B and C look like this:

$$A = w_A.a_1a_2\ldots a_{j-1}a_j\ldots a_{k-1}a_k\ldots$$

$$B = w_A.a_1a_2\ldots a_{j-1}b_j\ldots b_{k-1}b_k\ldots$$

$$C = w_A.a_1a_2\ldots a_{j-1}b_j\ldots b_{k-1}c_k\ldots$$

(again, do you see why?) Then A and C agree up to the jth place, but $a_j > b_j = c_j$, so $A > C$.

Finally, if $k < j$, A, B and C look like this:

$$A = w_A.a_1a_2\ldots a_{k-1}a_k\ldots a_{j-1}a_j\ldots$$

$$B = w_A.a_1a_2\ldots a_{k-1}a_k\ldots a_{j-1}b_j\ldots$$

$$C = w_A.a_1a_2\ldots a_{k-1}c_k\ldots c_{j-1}c_j\ldots$$

(Why?) Then A and C agree up to the kth place, but $a_k = b_k > c_k$, so again $A > C$.

This concludes the proof. $\square$

Definition 7.7.3 We write $A \geq B$ if $A > B$ or $A = B$. Likewise we write $A \leq B$ if $A < B$ or $A = B$.

For example, $3.14159265\ldots \geq 3.109635\ldots$ and likewise $3.14159265\ldots \geq 3.14159265\ldots$.

Next we show that as far as the ordering relations are concerned, finite decimals and their associated terminating decimals behave alike. Recall that if d is a finite decimal, d^* is its associated terminating decimal.

Theorem 7.7.4 *Let d_1 and d_2 be base n finite decimals. Then:*

1. *$d_1 > d_2$ if and only if $d_1^* > d_2^*$*

2. *$d_1 = d_2$ if and only if $d_1^* = d_2^*$.*

Proof Part 1) is immediate from the discussion in Section 6.5 on the ordering of finite decimals, and our definition of order for infinite decimals.

Part 2) follows easily from part 1). Suppose $d_1 = d_2$ but $d_1^* > d_2^*$ or $d_2^* > d_1^*$, say the first. Then by part 1) $d_1 > d_2$, contradicting the fact that $d_1 = d_2$. Thus if $d_1 = d_2$, we must have $d_1^* = d_2^*$. The proof of the converse is similar. □

Between any two rational numbers there is a finite decimal, as you were asked to show in Exercise 6.4.2. The next theorem, essentially, is a powerful generalization of this.

Theorem 7.7.5 (Denseness) *In base n, if A and B are infinite decimals with $A > B$, there is a terminating decimal T with $A > T > B$. Briefly, between any two infinite decimals there is a terminating decimal.*

Proof For convenience, we give the proof for base 10.

We may suppose A and B are in standard form, and as usual, we write:

$$A = w_A.a_1a_2a_3a_4\ldots$$

$$B = w_B.b_1b_2b_3b_4\ldots$$

Now $A > B$, and let us say this happens because A and B are the same out to the nth decimal place, but there A has a bigger digit than B has. Then our decimals really look like this:

$$A = w_A.a_1a_2\ldots a_{n-1}a_na_{n+1}\ldots$$

$$B = w_A.a_1a_2\ldots a_{n-1}b_nb_{n+1}\ldots$$

and $a_n > b_n$.

Next, not all of b_{n+1}, b_{n+2}, b_{n+3}, etc., can be nines, since then B would not be in standard form. So, in B, some decimal place after the nth must not be a 9, say the kth. Then

$$B = w_A.a_1a_2\ldots a_{n-1}b_nb_{n+1}\ldots b_{k-1}b_kb_{k+1}\ldots$$

where $b_k \neq 9$, so of course $9 > b_k$.

Now consider the terminating decimal

$$T = w_A.a_1a_2 \ldots a_{n-1}b_nb_{n+1} \ldots b_{k-1}90000\ldots$$

That is, we have changed the kth decimal place of B to a 9, and all later ones to 0's. We claim $A > T > B$.

First,

$$A = w_A.a_1a_2 \ldots a_{n-1}a_na_{n+1} \ldots a_{k-1}a_ka_{k+1} \ldots$$

$$T = w_A.a_1a_2 \ldots a_{n-1}b_nb_{n+1} \ldots b_{k-1}9000\ldots$$

and $A > T$ because $a_n > b_n$.

Next,

$$T = w_A.a_1a_2 \ldots a_{n-1}b_nb_{n+1} \ldots b_{k-1}9000\ldots$$

$$B = w_A.a_1a_2 \ldots a_{n-1}b_nb_{n+1} \ldots b_{k-1}b_kb_{k+1} \ldots$$

and $T > B$ because $9 > b_k$.

This concludes the proof. ⊏

This theorem has a corollary of great use which says, roughly, an infinite decimal is specified completely by saying what finite decimals are smaller than it. It will enable us to use facts about finite decimals to prove things about infinite decimals.

Theorem 7.7.6 (Equality) *In base n, let A and B be infinite decimals, and suppose that for each terminating decimal T, $A > T$ precisely when $B > T$; then $A = B$.*

Proof If $A \neq B$, by the trichotomy law one must be bigger; say $A > B$. Then there is a terminating decimal T such that $A > T > B$. But then there is a terminating decimal T with $A > T$ but without $B > T$. □

Exercises

Exercise 7.7.1 Rank the following base 5 infinite decimals in order of size from biggest to smallest:

2.1342401...

31.42312...

2.140000...

31.42203...

2.13444...

Exercise 7.7.2 Complete the proof of Theorem 7.7.1 by showing we can't have $A = B$ and also $A > B$.

Exercise 7.7.3 Prove the following:

1. if $A > B$ and $B \geq C$ then $A > C$

2. if $A \geq B$ and $B > C$ then $A > C$

3. if $A \geq B$ and $B \geq C$ then $A \geq C$

Exercise 7.7.4 Let $A = w_A.a_1a_2a_3a_4\ldots$ and $B = w_B.b_1b_2b_3b_4\ldots$ be base n infinite decimals in standard form. Show $A \geq B$ if and only if $w_A \geq w_B$ and $w_A.a_1 \geq w_B.b_1$ and $w_A.a_1a_2 \geq w_B.b_1b_2$ etc.

Exercise 7.7.5 Use the method of the proof of Theorem 7.7.5 and produce a terminating decimal between $1.236418926\ldots$ and $1.236299783\ldots$.

Exercise 7.7.6 Using the same set-up as in the proof of Theorem 7.7.5, suppose $A > B$ because $w_A > w_B$ and find a terminating decimal between A and B.

Exercise 7.7.7 Carry out a proof similar to that of Theorem 7.7.5 for base 2.

Exercise 7.7.8 Carry out the proof of Theorem 7.7.5 in full generality, for base n.

Exercise 7.7.9 Suppose that for each terminating decimal T, $B > T$ implies $A > T$. Prove $A \geq B$.

7.8   Addition – an introduction

We want to define addition of infinite decimals so that it agrees with our intuitions in these matters. Let us try to add two infinite decimals, then see what is necessary to justify what we would like to do. Consider the base 10 sum

$$2.43271006\ldots$$
$$+ \quad 1.46829102\ldots$$

Since these are infinite decimals we can't start at the right end — there is none. This is the origin of much that is peculiar about infinite decimals. On the other hand, if we try adding from left to right we get into the problem of carrying. Still, we can add more and more places of them. For instance, after adding two places we have

$$2.43271006\ldots$$
$$+ \quad 1.46829102\ldots$$
$$\overline{\quad 3.89}$$

and, after adding four places we have

$$
\begin{array}{r}
2.43271006\ldots \\
+ \quad 1.46829102\ldots \\
\hline
3.9009
\end{array}
$$

We can easily compute the sums of as many places as we like.

Consider the following sequence of partial sums, arising from this problem:

$2 + 1$	$=$	3
$2.4 + 1.4$	$=$	3.8
$2.43 + 1.46$	$=$	3.89
$2.432 + 1.468$	$=$	3.900
$2.4327 + 1.4682$	$=$	3.9009
$2.43271 + 1.46829$	$=$	3.90100
$2.432710 + 1.468291$	$=$	3.901001
$2.4327100 + 1.4682910$	$=$	3.9010010
$2.43271006 + 1.46829102$	$=$	3.90100108
etc.		etc.

Is there some reasonable sense in which the sequence of finite decimals thus produced can be said to 'tend to' some infinite decimal which we could then call the sum?

Notice, in the sequence above, the whole number part is always 3. The first decimal place begins as 8, then becomes a 9 and apparently remains 9. The second decimal place begins as 9, then becomes and seems to stay 0. And so on. Suppose we could show that in this sequence, after some initial number of steps, any given decimal place in fact becomes fixed. Then it would be reasonable to say the sequence tends to the infinite decimal in which, say, the third decimal place is the digit the third decimal places of our partial sums eventually fix on; similarly for the fourth decimal place, and so on. In this sense the sequence above would tend to an infinite decimal beginning 3.901001....

The sequence above of partial sums is *non-decreasing*, that is, each term is at least as big as the one before it. It is easy to see this must be the case for any two infinite decimals we try to add. Also, each term is smaller than 4. In fact, these are the only properties we need to guarantee the sequence will 'tend to' an infinite decimal limit in the sense we have described.

7.9 The basic theorems on limits

In this section we establish the results needed to add infinite decimals in the way suggested in the previous section. What we do here is of fundamental importance and will be used many times in the rest of the book.

Definition 7.9.1 Let $A_1, A_2, A_3, A_4, \ldots$ be an infinite sequence of base n infinite decimals. We say the sequence is *non-decreasing* if $A_1 \leq A_2 \leq A_3 \leq A_4 \leq \ldots$. We say the sequence is *bounded from above* if there is a base n infinite decimal D such that $A_1 \leq D$, $A_2 \leq D$, $A_3 \leq D$, etc.

Theorem 7.9.2 (The Limit Theorem)
Let $A_1, A_2, A_3, A_4, \ldots$ be an infinite sequence of base n infinite decimals all in standard form, which is non-decreasing and bounded from above. From some point on in this sequence the whole number parts are all the same; from some point on in this sequence the whole number parts and first decimal places are all the same; similarly for the whole number parts, first and second decimal places; and so on.

Proof Let us say the infinite decimals look like this:

$$A_1 = w_1.a_{11}a_{12}a_{13}\ldots$$

$$A_2 = w_2.a_{21}a_{22}a_{23}\ldots$$

$$A_3 = w_3.a_{31}a_{32}a_{33}\ldots$$

$$\text{etc.}$$

and let us say D bounds the sequence from above, where

$$D = w_D.d_1d_2d_3d_4\ldots$$

In these, $w_1, w_2, \ldots$ and w_D are whole number names in base n notation, and each a_{ij} and d_j is a base n digit. (Since we have an infinite sequence of decimals, each of which involves an infinite sequence of decimal places, we have had to resort to double subscripts.)

D bounds the sequence from above, that is, $A_1 \leq D$, $A_2 \leq D$, $A_3 \leq D$, etc., so it must be that $w_1 \leq w_D$, $w_2 \leq w_D$, $w_3 \leq w_D$, etc. (Why?) Then there are only a finite number of different whole number parts for all of A_1, A_2, A_3, etc. (there are not more than w_D possibilities). So among them there is a largest (see Theorem 2.13.5.) Let us say this largest whole number part occurs first in the qth infinite decimal in our sequence, that is, in

$$A_q = w_q.a_{q1}a_{q2}a_{q3}\ldots$$

We claim from this point on in the sequence the whole number parts will all be the same, for the following reasons. No whole number part can be bigger than w_q, since that was chosen to be biggest of all the whole number parts. And if A_k occurs later in the sequence than A_q, its whole number part can't be smaller than that of A_q, since if it was, that would make A_k smaller than A_q, $A_k < A_q$, but our sequence is non-decreasing, and A_k comes after A_q, so $A_q \leq A_k$.

Thus from A_q on, all the infinite decimals in our sequence have the same whole number parts.

Now we repeat the argument to fix the first decimal place, but starting with A_q instead of at the beginning, A_1.

There are only n base n digits, so among the digits that occur in the first decimal places of the terms in the sequence A_q, A_{q+1}, A_{q+2}, ... there must be a biggest. Let us say the first term in our sequence starting from A_q which has this biggest digit in its first decimal place is A_r, where

$$A_r = w_r.a_{r1}a_{r2}a_{r3}\ldots$$

We claim from this point on in the sequence, all infinite decimals will have the same first decimal place too, for these reasons. First, no infinite decimal after A_q can have a bigger digit in its first decimal place than a_{r1}, since that was chosen to be biggest of all. And if A_k occurs later in the sequence than A_r its first decimal place can't be smaller than that of A_r because that would make A_k smaller than A_r (both are later in the sequence than A_q, so both have the same whole number parts), $A_k < A_r$, but the sequence is non-decreasing, so $A_r \leq A_k$.

Thus from A_r on, all the infinite decimals in our sequence have the same first decimal places as well as the same whole number parts.

Next, the argument can be repeated to fix the second decimal place, then the third, the forth, and so on. (Strictly speaking, there is a proof using induction involved here.) □

Definition 7.9.3 Let $A_1, A_2, A_3, A_4, \ldots$ be the sequence of the theorem above. There is an infinite decimal that may naturally be associated with the sequence: that infinite decimal whose whole number part is what the whole number parts of terms of the sequence eventually settle on, whose first decimal place is the digit that the first decimal places of terms of the sequence eventually settle on, and so on. We call this the infinite decimal *generated by* the sequence.

Let L be the infinite decimal generated by the sequence. If L is in standard form we also say it is the *limit* of the sequence; if L is not in standard form, we call the result of putting it in standard form the limit.

In these terms, we have shown

Corollary 7.9.4 *A non-decreasing sequence of base n infinite decimals (in standard form) which is bounded from above has a limit.*

Example Consider the non-decreasing bounded sequence of base 10 infinite decimals,

$$A_1 = 21.326812\ldots$$
$$A_2 = 22.826134\ldots$$
$$A_3 = 22.828268\ldots$$
$$A_4 = 22.837192\ldots$$
$$A_5 = 22.837293\ldots$$
$$\text{etc.}$$

Let us suppose that the whole number parts are all 22 from A_2 on, that the first decimal places are all 8's from A_2 on, that the second decimal places are all 3's from A_4 on, and the third decimal places are all 7's from A_4 on. Then the infinite decimal generated by this sequence begins 22.837..., and all we can say for sure, given this information, is that the limit begins either 22.837... or 22.838... (Why?)

Example Consider the non-decreasing bounded sequence of base 10 infinite decimals

$$A_1 = 1.826315\ldots$$
$$A_2 = 1.937182\ldots$$
$$A_3 = 1.992813\ldots$$
$$A_4 = 1.999712\ldots$$
$$\text{etc.}$$

Suppose the whole number parts are all 1's, and each decimal place eventually settles on 9. Then the infinite decimal generated by this sequence is 1.9999..., and the limit is 2.000....

We remarked in Section 2 that non-constructive proofs were inherent in the theory of real numbers. The proof above is one, yet it is essential to the development of the theory. Recall, using the terminology of the proof above, w_q was the largest whole number part of any infinite decimal in the sequence $A_1, A_2, A_3, A_4, \ldots$. But to determine the exact size of w_q we have to 'look at' all the terms of the sequence to see what the biggest whole number part actually is, and this process can't be carried out. We may accept that such a number must exist, but we may not be able to determine it for a specific sequence. There is a similar difficulty in knowing whether the infinite decimal generated by the sequence is in standard form or not. (Do you see why?) Yet we may assume either it is or it isn't, even though we may not know which.

The limit we produced for the sequence $A_1, A_2, A_3, A_4, \ldots$ in the proof above has some interesing properties. One is that the terms of the sequence get arbitrarily close to that limit. We can't establish this yet, as we have no definition of subtraction, and hence of closeness. There is another property that we can discuss now, however.

Definition 7.9.5 Let **C** be a collection of base n infinite decimals, and let L be a base n infinite decimal. L is called a *least upper bound* for **C** if:

1. L is an *upper bound* for **C**, that is, for each infinite decimal A in **C**, $A \leq L$,

2. no infinite decimal smaller than L is an upper bound for **C**; that is, if L' is also an upper bound for **C** we don't have $L' < L$, or equivalently, we must have $L \leq L'$.

Theorem 7.9.6 *Let $A_1, A_2, A_3, A_4, \ldots$ meet the hypotheses of the limit theorem. The limit of this sequence is also the least upper bound of the collection $\{A_1, A_2, A_3, A_4, \ldots\}$.*

Proof Let L be the infinite decimal generated by the sequence $A_1, A_2, A_3, \ldots$.

Case 1) L is in standard form. Then L itself is the limit. It should be clear from the construction of L that the following is the case. The whole number part of L is as big as the whole number part of any term in the sequence. The first decimal place of L is as big as the first decimal place of any term of the sequence having the same whole number part as L. Similarly for the second decimal place, the third, and so on. Then L must be as big as any term of the sequence, that is, L is an *upper bound*.

Next, suppose there is a smaller upper bound L', which we may assume is in standard form. $L' < L$, and for the sake of argument, let us say this happens because L and L' have the same whole number parts, the same first decimal places, but L' has a smaller second decimal place than L. But again, consider the way L was constructed. From some point on in the sequence $A_1, A_2, A_3, A_4, \ldots$, all terms have the same whole number part, and first and second decimal paces as L. Let us say this happens from A_s onward. Then, in particular, A_s agrees with L through the second decimal place. But L' has the same whole number part as L, the same first decimal place, but a smaller second decimal place, so this makes L' smaller than A_s; so L' couldn't have been an upper bound after all. Thus L is a *least* upper bound.

Case 2) L is not in standard form. Then the limit is the result of putting L into standard form, let us call it L_S. Rather than work with this case in detail, we give you Exercise 7.9.3; doing it will illustrate the ideas on which a proof in Case 2 can be based. □

Not every collection of infinite decimals has a least upper bound. Consider, for example, the collection of base 10 terminating decimals $\{(1)^*, (2)^*, (3)^*, (4)^*, \ldots\}$. What we do know from our work so far is that if a collection of infinite decimals can be arranged into a non-decreasing sequence which is bounded from above, then it has a least upper bound. But in fact, least upper bounds exist under more general conditions than this. The following theorem shows this, with a proof very similar to those we have seen so far. It is a central fact about real numbers.

Theorem 7.9.7 (The Least Upper Bound Theorem)
Any (non-empty) collection of base n infinite decimals which has an upper bound has a least upper bound.

Proof Let $\mathbf{C}$ be a collection of base n infinite decimals, which we may suppose have been put into standard form if necessary, and suppose $\mathbf{C}$ has $D = w_D.d_1 d_2 d_3 d_4 \ldots$ as an upper bound.

No member of $\mathbf{C}$ can have a larger whole number part than w_D so there are only a finite number of different whole number parts for members in $\mathbf{C}$. Let w_L be the largest one.

Next, there are only n base n digits, so there must be a largest digit which occurs as first decimal place of an infinite decimal in $\mathbf{C}$ with w_L as its whole number parts; say it is l_1.

Similarly there must be a largest digit which occurs as second decimal place in those members of $\mathbf{C}$ beginning with $w_L.l_1$. Say it is l_2.

In a similar fashion we produce l_3, l_4, l_5, etc. Let $L = w_L.l_1l_2l_3l_4\ldots$. We claim L is the least upper bound of $\mathbf{C}$. There are two cases, depending on whether or not L is in standard form. We leave the rest to you in Exercises 7.9.4 and 7.9.5. □

The Least Upper Bound Theorem is characteristic of the real number system. In the rational number system, for example, we can form the collection $\{1, 1.4, 1.41, 1.41, \ldots\}$ of better and better approximations to $\sqrt{2}$. We will see later that the only possible least upper bound for this collection is $\sqrt{2}$ itself, and we know that is not a rational number.

Finally, a result that will be of much use in establishing properties of 'elementary' arithmetic for infinite decimals.

Theorem 7.9.8 (The inequality theorem)
In base n, let $A_1, A_2, A_3, A_4, \ldots$ and $B_1, B_2, B_3, B_4, \ldots$ be two sequences of infinite decimals in standard form, both non-decreasing and both bounded from above. Let the limit of $A_1, A_2, A_3, \ldots$ be A, and the limit of $B_1, B_2, B_3, \ldots$ be B. Suppose $A_1 \geq B_1$, $A_2 \geq B_2$, $A_3 \geq B_3$, etc. Then $A \geq B$.

Proof Suppose we don't have $A \geq B$, so we do have $A < B$. Now B is the least upper bound of $\{B_1, B_2, B_3, \ldots\}$ and by Exercise 7.9.2, if $A \geq B_1$, $A \geq B_2$, $A \geq B_3$, etc., we would have $A \geq B$. So it must be that for some k we don't have $A \geq B_k$, so we have $A < B_k$. But A is the least upper bound of $\{A_1, A_2, A_3, \ldots\}$, so $A \geq A_k$, and by hypothesis, $A_k \geq B_k$, hence $A \geq B_k$. This is a contradiction, and concludes the proof. □

Exercises

Exercise 7.9.1 Prove that no collection $\mathbf{C}$ of base n infinite decimals can have more than one least upper bound.

Exercise 7.9.2 Let $\mathbf{C}$ be a collection of infinite decimals, with L as a least upper bound. Suppose A is bigger than or equal to every member of $\mathbf{C}$. Show $A \geq L$.

Exercise 7.9.3 Suppose we are working in base 10. Say the infinite decimal generated by the sequence $A_1, A_2, A_3, \ldots$ is $L = 1.2399999\ldots$, so that the limit is $L_S = 1.240000\ldots$.

1. Prove L_S is an upper bound for the sequence $A_1, A_2, A_3, \ldots$.

2. Prove that if $L' < L_S$, then L' must be smaller than some term of the sequence, and hence not an upper bound. Thus L_S is a least upper bound.

Exercise 7.9.4 Assume L, as given in the proof of Theorem 7.9.7, is in standard form and prove it is the least upper bound for **C**.

Exercise 7.9.5 Assume L, as given in the proof of Theorem 7.9.7, is not in standard form. For simplicity suppose we are working in base 10, and $L = 1.239999\ldots$. Then the result of putting L in standard form is $1.240000\ldots$. Show this is the least upper bound for **C**. See Exercise 7.9.3. Note: this is not really a proof of case 2, but it will amply illustrate the ideas of such a proof.

Exercise 7.9.6 Use the fact that $\frac{1}{3}$ is not a base 10 finite decimal to construct another example showing the Least Upper Bound Theorem does not hold for the system of base 10 finite decimals.

7.10 Addition of infinite decimals

We now have developed enough material to carry out the ideas of Section 7.8 rigorously. Suppose we want to add in base 10

$$38.263527\ldots$$
$$+ \quad 18.172938\ldots$$

First we can compute using *finite decimals* the partial sums:

$$
\begin{array}{lcl}
38 + 18 & = & 56 \\
38.2 + 18.1 & = & 56.3 \\
38.26 + 18.17 & = & 56.43 \\
38.263 + 18.172 & = & 56.435 \\
38.2635 + 18.1729 & = & 56.4364 \\
\text{etc.} & & \text{etc.}
\end{array}
$$

The sequence of finite decimals we produce is clearly non-decreasing (since we are including more and more places). If we replace each finite decimal with its associated terminating decimal we get the following non-decreasing sequence of infinite decimals:

$$56.00000000\ldots$$
$$56.30000000\ldots$$
$$56.43000000\ldots$$
$$56.43500000\ldots$$
$$56.43640000\ldots$$
$$\text{etc.}$$

Since all these are terminating decimals they are all in standard form. Also, the sequence has $(39)^* + (19)^* = (58)^* = 58.0000\ldots$ as an upper bound.

By the Limit Theorem 7.9.2 this sequence has a limit, which we *define* to be the sum of $38.263527\ldots$ and $18.172938\ldots$.

More generally, let A and B be base n infinite decimals, say in standard form they are

$$A = w_A.a_1a_2a_3a_4\ldots$$

and

$$B = w_B.b_1b_2b_3b_4\ldots$$

First, compute using base n finite decimals:

$$
\begin{aligned}
w_A + w_B &= d_0 \\
w_A.a_1 + w_B.b_1 &= d_1 \\
w_A.a_1a_2 + w_B.b_1b_2 &= d_2 \\
w_A.a_1a_2a_3 + w_B.b_1b_2b_3 &= d_3 \\
\text{etc.} & \quad \text{etc.}
\end{aligned}
$$

Form the associated sequence of terminating decimals, $d_0^*, d_1^*, d_2^*, d_3^*, \ldots$ It should be clear that this is non-decreasing. But also, as base n finite decimals

$$
\begin{aligned}
d_k &= w_A.a_1a_2a_3\ldots a_k + w_B.b_1b_2b_3\ldots b_k \\
&\leq (w_A + 1) + (w_B + 1)
\end{aligned}
$$

and so, for each k,

$$(d_k)^* \leq [w_A + w_B + 2]^*$$

thus we have a sequence that is bounded from above, and so, by the Limit Theorem, $d_0^*, d_1^*, d_2^*, d_3^*, \ldots$ has a limit.

Definition 7.10.1 We define $A + B$ to be the limit of the sequence of terminating decimals $d_0^*, d_1^*, d_2^*, d_3^*, \ldots$ constructed above.

Exercises

Exercise 7.10.1 Use the definition and show, in base 10,

$$
\begin{array}{r}
.33333\ldots \\
+ \quad .33333\ldots \\
\hline
.66666\ldots
\end{array}
$$

Exercise 7.10.2 Show in base 10,

$$
\begin{array}{r}
.33333\ldots \\
+ \quad .66666\ldots \\
\hline
1.00000\ldots
\end{array}
$$

Exercise 7.10.3 Show in base 2,

$$
\begin{array}{r}
.010101\ldots \\
+ \quad .101010\ldots \\
\hline
1.000000\ldots
\end{array}
$$

Exercise 7.10.4 In base 10,

$$
\begin{array}{r}
.14 \\
+ \quad .28 \\
\hline
.42
\end{array}
$$

Now use the definition above to show

$$
\begin{array}{r}
.1400000\ldots \\
+ \quad .2800000\ldots \\
\hline
.4200000\ldots
\end{array}
$$

Remark The previous exercise says, in base 10, $(.14)^* + (.28)^* = (.14 + .28)^*$. It is a special case of a general result which says finite decimals and their associated terminating decimals behave alike with respect to addition.

Exercise 7.10.5 In base n, let d_1 and d_2 be finite decimals. Show $d_1^* + d_2^* = (d_1 + d_2)^*$. Suggestion: write d_1 as $w_1.a_1a_2\ldots a_k$ and d_2 as $w_2.b_1b_2\ldots b_k$ and also $d_1 + d_2$ as $w_3.c_1c_2\ldots c_k$ then construct the 'partial sum' sequence for adding d_1^* and d_2^*, beginning at the $k+1^{st}$ term (the first k terms won't affect the limit at all).

7.11 Basic properties of addition

As with the other number systems, addition here is commutative and associative (hence a sum of more than two items can be rearranged), and addition interacts in useful ways with the ordering relation. Commutativity is easy to show and we leave the work to you, but associativity is more difficult. Much of this section is spent proving it.

Theorem 7.11.1 (Commutativity of addition)
If A and B are base n infinite decimals then $A + B = B + A$.

Theorem 7.11.2 *Let A, B and C be base n infinite decimals. If $A \geq B$ then $A + C \geq B + C$.*

Proof We may suppose A, B and C are in standard form. Let

$$
\begin{aligned}
A &= w_A.a_1a_2a_3a_4\ldots \\
B &= w_B.b_1b_2b_3b_4\ldots \\
C &= w_C.c_1c_2c_3c_4\ldots
\end{aligned}
$$

$A + C$ is the limit of the sequence

$$
\begin{aligned}
&(w_A + w_C)^* \\
&(w_A.a_1 + w_C.c_1)^* \\
&(w_A.a_1a_2 + w_C.c_1c_2)^* \\
&\text{etc.}
\end{aligned}
$$

and $B + C$ is the limit of the sequence

$$(w_B + w_C)^*$$
$$(w_B.b_1 + w_C.c_1)^*$$
$$(w_B.b_1b_2 + w_C.c_1c_2)^*$$
etc.

$A \geq B$ so we get (Exercise 7.7.4)

$$
\begin{array}{ccc}
w_A & \geq & w_B \\
w_A.a_1 & \geq & w_B.b_1 \\
w_A.a_1a_2 & \geq & w_B.b_1b_2
\end{array}
$$
etc.

Now, by already established properties of *rational numbers* (which includes the *finite* decimals)

$$
\begin{array}{ccc}
w_A + w_C & \geq & w_B + w_C \\
w_A.a_1 + w_C.c_1 & \geq & w_B.b_1 + w_C.c_1 \\
w_A.a_1a_2 + w_C.c_1c_2 & \geq & w_B.b_1b_2 + w_C.c_1c_2
\end{array}
$$
etc.

Since terminating decimals behave like their corresponding finite decimals with respect to order (Theorem 7.7.4) we have

$$
\begin{array}{ccc}
(w_A + w_C)^* & \geq & (w_B + w_C)^* \\
(w_A.a_1 + w_C.c_1)^* & \geq & (w_B.b_1 + w_C.c_1)^* \\
(w_A.a_1a_2 + w_C.c_1c_2)^* & \geq & (w_B.b_1b_2 + w_C.c_1c_2)^*
\end{array}
$$
etc.

Now an application of the Inequality Theorem 7.9.8 gives us

$$A + C \geq B + C.$$

□

Our next main theorem is the associativity of addition. But first we need a few preliminary results.

Lemma 7.11.3 *In base n, suppose U, V and T are finite decimals, with $U \neq 0$ and $V \neq 0$. Suppose also that $U + V > T$. Then there are finite decimals a and b such that $U > a$, $V > b$, and $a + b > T$.*

Proof We have three cases.

Case 1) $T \geq U$ and $T \geq V$. Since $T \geq V$, $T - V$ is defined in the system of rational numbers. Now, $U + V > T$ so $U > T - V$. Let a be a base n finite decimal between U and $T - V$ (Exercise 6.4.2). Then $U > a$, and $a > T - V$ so $a + V > T$. Also $T \geq U > a$, so $T - a$ is defined in the rationals. Now,

$a + V > T$ so $V > T - a$. Let b be a base n finite decimal between V and $T - a$. Then $V > b$, and $b > T - a$ so $a + b > T$. This completes case 1, the most complicated.

Case 2) $T < U$. In this case, take b to be 0. Since $V \neq 0$, $V > b$. And take a to be any base n finite decimal between T and U. Then $U > a$, and $a + b = a > T$.

Case 3) $T < V$ is treated similarly to case 2). $\square$

Notice that since finite decimals and terminating decimals behave alike with respect to order and addition, we immediately have a version of this lemma for terminating decimals too.

Now the key lemma which will allow us to use the Equality Theorem of Section 7.7.

Lemma 7.11.4 (Key Lemma) *In base n, let $A \neq 0^*$ and $B \neq 0^*$ be infinite decimals and T be a terminating decimal. Suppose $A + B > T$. Then there are terminating decimals a and b such that $A > a$, $B > b$ and $a + b > T$.*

Proof We may suppose A and B are in standard form. Let

$$A = w_A.a_1 a_2 a_3 a_4 \ldots$$
$$B = w_B.b_1 b_2 b_3 b_4 \ldots$$

Now $A + B$ is the limit of the sequence

$$L_0 = (w_A + w_B)^*$$
$$L_1 = (w_A.a_1 + w_B.b_1)^*$$
$$L_2 = (w_A.a_1 a_2 + w_B.b_1 b_2)^*$$
$$\text{etc.}$$

and hence also, as we showed in Section 7.9, $A + B$ is the least upper bound of this sequence. $A + B > T$, so T is not an upper bound, so for some j we don't have $T \geq L_j$; then we do have $L_j > T$.

$A \neq 0^*$, so not all of w_A, a_1, a_2, $a_3 \ldots$ can be 0. The same applies to B. Choose any $k \geq j$ so that among w_A, $a_1, \ldots, a_k$ is a non-zero term, and among w_B, $b_1, \ldots, b_k$ is a non-zero term.

Since $k \geq j$, $L_k > T$ (since $L_j > T$ and the sequence is non-decreasing). That is, $(w_A.a_1 a_2 \ldots a_k + w_B.b_1 b_2 \ldots b_k)^* > T$. But since finite and terminating decimals behave alike with respect to addition, this says

$$(w_A.a_1 a_2 \ldots a_k)^* + (w_B.b_1 b_2 \ldots b_k)^* > T.$$

Set

$$U = (w_A.a_1 a_2 \ldots a_k)^*$$
$$V = (w_B.b_1 b_2 \ldots b_k)^*$$

Then we have arranged things so that U and V are terminating decimals, neither 0^*, and $U + V > T$. Also, clearly, $A \geq U$ and $B \geq V$ (why?).

Now by the previous lemma (transferred to terminating decimals) there are terminating decimals a and b with $U > a$, $V > b$ and $a + b > T$. Since $A \geq U > a$, $A > a$; similarly $B > b$. This concludes the proof. $\square$

Finally we are ready to show

Theorem 7.11.5 (Associativity of Addition)
In base n, let A, B and C be infinite decimals. $A + (B + C) = (A + B) + C$.

Proof If any of A, B or C is 0^*, the result is immediate using Exercise 7.11.4. So now we suppose none is 0^*.

By the Equality Theorem 7.7.6 it is enough to show that for each terminating decimal T, $A + (B + C) > T$ if and only if $(A + B) + C > T$. We show the implication one way only, the other direction being similar.

Well, let T be a terminating decimal, and suppose $A + (B + C) > T$. By the Key Lemma 7.11.4, there are terminating decimals a and d such that $A > a$, $B + C > d$ and $a + d > T$. Now $B + C > d$, so by the Key Lemma again there are terminating decimals b and c such that $B > b$, $C > c$ and $b + c > d$. Now, $A \geq a$, $B \geq b$ and $C \geq c$, so by Exercise 7.11.2 $A + B \geq a + b$ and $(A + B) + C \geq (a + b) + c$. But terminating decimal addition is associative (Exercise 7.11.5). Thus $(a + b) + c = a + (b + c)$ but also, $b + c \geq d$, so $a + (b + c) \geq a + d > T$ so $(A + B) + C > T$. This ends the proof. $\square$

We conclude this section with some results about our definition of addition. Let A and B be base n infinite decimals in standard form; as usual, set

$$A = w_A.a_1a_2a_3a_4\ldots$$
$$B = w_B.b_1b_2b_3b_4\ldots$$

Now, $A + B$ was defined to be the limit of the sequence

$$(w_A + w_B)^*$$
$$(w_A.a_1 + w_B.b_1)^*$$
$$(w_A.a_1a_2 + w_B.b_1b_2)^*$$
etc.

In this sequence, each term is essentially a sum of an approximation to A and an approximation to B, in each case, approximations to the same number of places. Is there something important about having the approximations to both A and B of the same length, or it is incidental. For example, would the following sequence also have $A + B$ as a limit:

$$(w_A + w_B)^*$$
$$(w_A.a_1a_2 + w_B.b_1)^*$$
$$(w_A.a_1a_2a_3a_4 + w_B.b_1b_2)^*$$
$$(w_A.a_1a_2a_3a_4a_5a_6 + w_B.b_1b_2b_3)^*$$
etc.

In this the A approximations are twice as long as the B approximations.

Theorem 7.11.6 *In base n, let A and B be infinite decimals. Let $\mathbf{C}_A$ be any collection of infinite decimals having A as least upper bound, and let $\mathbf{C}_B$ be any collection having B as least upper bound. Let $\mathbf{C}$ consist of all sums of terms from $\mathbf{C}_A$ with terms from $\mathbf{C}_B$. That is,*

$$\mathbf{C} = \{x + y \mid x \in \mathbf{C}_A \text{ and } y \in \mathbf{C}_B\}.$$

Then $\mathbf{C}$ has $A + B$ as least upper bound.

Proof If $A = 0^*$, then $\mathbf{C} = \mathbf{C}_B$ (why?) so the least upper bound of $\mathbf{C}$ is that of $\mathbf{C}_B$, namely $B = 0^* + B = A + B$. Similarly if $B = 0^*$.

Now suppose neither A nor B is 0^*. We first show $A + B$ is an upper bound for $\mathbf{C}$. Suppose $z \in \mathbf{C}$. Then for some $x \in \mathbf{C}_A$ and $y \in \mathbf{C}_B$, $z = x + y$. Since $x \in \mathbf{C}_A$, and A is the least upper bound of $\mathbf{C}_A$, $x \leq A$. Similarly $y \leq B$. Then $z = x + y \leq A + B$. Thus $A + B$ is an upper bound for $\mathbf{C}$.

Since $\mathbf{C}$ has an upper bound, it has a *least* upper bound, call it C. We immediately have $C \leq A + B$.

Suppose $C < A + B$. Then there is a terminating decimal T between them, $C < T < A + B$. Since $A + B > T$, by the Key Lemma 7.11.4, there are terminating decimals a and b such that $A > a$, $B > b$, and $a + b > T$. Since A is the least upper bound of $\mathbf{C}_A$ and $A > a$, a is not an upper bound for $\mathbf{C}_A$. Then there must be an a' in $\mathbf{C}_A$ with $a' \geq a$. Similarly there must be some b' in $\mathbf{C}_B$ with $b' \geq b$. Now, $a' + b'$ is in $\mathbf{C}$ since a' is in $\mathbf{C}_A$ and b' is in $\mathbf{C}_B$. But, $a' + b' \geq a + b > T > C$ and C is the least upper bound of $\mathbf{C}$. This is a contradiction. We can't have $C < A + B$, so we must have $C = A + B$.

This concludes the proof. □

Now we can easily answer the question raised just before this theorem.

Corollary 7.11.7 *In base n, let A_1, A_2, A_3, $A_4, \ldots$ be any bounded, non-decreasing sequence of infinite decimals with A as a limit, and let B_1, B_2, B_3, $B_4, \ldots$ be any bounded non-decreasing sequence with B as a limit. Then the sequence $A_1 + B_1$, $A_2 + B_2$, $A_3 + B_3, \ldots$ has $A + B$ as a limit.*

One important consequence of this corollary is the fact that in adding A and B it doesn't matter whether we use standard forms or non-standard ones. We illustrate this with a base 10 example. Consider the problem $A + B$ where $A = (3)^*$ and $B = (2)^*$.

$$
\begin{array}{r}
3.0000\ldots \\
+\quad 2.0000\ldots \\
\hline
5.0000\ldots
\end{array}
$$

Suppose we replace these by their non-standard forms, and apply the definition of addition even though this is technically incorrect.

$$
\begin{array}{r}
2.9999\ldots \\
+\quad 1.9999\ldots
\end{array}
$$

To work this out we form the sequence of partial sums:

$$
\begin{aligned}
L_1 &= (2.+1.)^* \\
L_2 &= (2.9+1.9)^* \\
L_3 &= (2.99+1.99)^*
\end{aligned}
$$

etc.

We claim this sequence has $5.0000\ldots$ as limit too.

Well, let $A_1 = (2)^*$, $A_2 = (2.9)^*$, $A_3 = (2.99)^*$, etc. Then the limit of this sequence is $A = (3)^*$. And let $B_1 = (1)^*$, $B_2 = (1.9)^*$, $B_3 = (1.99)^*$, etc. The limit of this sequence is $B = (2)^*$. By the corollary above, the sequence $A_1 + B_1$, $A_2 + B_2$, $A_3 + B_3$, etc. has $A + B = (3)^* + (2)^* = (5)^*$ as limit. But, $A_1 + B_1 = L_1$, $A_2 + B_2 = L_2$ and so on. So the above sequence of partial sums has $(5)^*$ as limit too.

Exercises

Exercise 7.11.1 Prove Theorem 7.11.1.

Exercise 7.11.2 Let A, B, C and D be base n infinite decimals. Suppose $A \geq C$ and $B \geq D$, and show $A + B \geq C + D$.

Exercise 7.11.3 The techniques of the proof of Lemma 7.11.4 may be used to show a result about non-standard forms. In base n, let A and B be infinite decimals in standard form, and suppose A also has a non-standard form, let $w_A.a_1a_2a_3a_4\ldots$ be it. If $A > B$, show that for some k, $(w_A.a_1a_2\ldots a_k)^* > B$. For example, in base 10, let $A = 3.1420000\ldots$ and $B = 3.1419962\ldots$. Then $A > B$. The non-standard form of A is $3.1419999\ldots$, and, in fact, $(3.141999)^* > B$.

Exercise 7.11.4 Prove: In base n, for any infinite decimal A, $A + 0^* = A$.

Exercise 7.11.5 In base n, let a, b and c be finite decimals, and show $a^* + (b^* + c^*) = (a^* + b^*) + c^*$. Hence addition of *terminating* decimals is associative.

Exercise 7.11.6 Give a proof for Corollary 7.11.7.

7.12   Multiplication of infinite decimals

Now that we have the pattern of addition to follow, we can treat multiplication easily. Suppose we wanted to give meaning to the base 10 problem:

$$
\begin{array}{r}
7.12368421\ldots \\
\times \quad 2.60827035\ldots \\
\hline
\end{array}
$$

Well, let us multiply out, as *finite* decimals, the *partial products*.

$$7 \times 2 \qquad\qquad = \quad 14.$$
$$7.1 \times 2.6 \qquad\quad = \quad 18.46$$
$$7.12 \times 2.60 \qquad = \quad 18.5120$$
$$7.123 \times 2.608 \quad = \quad 18.576784$$
$$7.1236 \times 2.6082 = \quad 18.57977352$$
$$\text{etc.} \qquad\qquad\qquad\quad \text{etc.}$$

We would like to say the limit of this sequence is the product of our two infinite decimals. But we can! The sequence must be non-decreasing since we are multiplying together longer and longer finite decimals. And the sequence is clearly bounded by $8 \times 3 = 24$. So

$$14.000000000\ldots$$
$$18.460000000\ldots$$
$$18.512000000\ldots$$
$$18.576784000\ldots$$
$$18.579773520\ldots$$
$$\text{etc.}$$

is a non-decreasing, bounded sequence of infinite decimals which, by the Limit Theorem 7.9.2 has a limit, apparently beginning $18.57\ldots$. And we can define the product to be this limit.

In general, let A and B be base n infinite decimals, say, in standard form,

$$A = w_A.a_1a_2a_3a_4\ldots$$

$$B = w_B.b_1b_2b_3b_4\ldots$$

Compute the finite decimal products

$$(w_A) \cdot (w_B)$$
$$(w_A.a_1) \cdot (w_B.b_1)$$
$$(w_A.a_1a_2 \cdot (w_B.b_1b_2)$$
$$\text{etc.}$$

This is a non-decreasing sequence of finite decimals. Also every term in the sequence is less than $(w_A + 1) \cdot (w_B + 1)$. Now consider the associated sequence of terminating decimals.

$$[(w_A) \cdot (w_B)]^*$$
$$[(w_A.a_1) \cdot (w_B.b_1)]^*$$
$$[(w_A.a_1a_2) \cdot (w_B.b_1b_2)]^*$$
$$\text{etc.}$$

This is a bounded, non-decreasing sequence of infinite decimals. Let L be the limit. *We define $A \cdot B = L$.*

Exercises

Exercise 7.12.1 In base 10, show $(10)^* \cdot (.3333\ldots) = 3.333\ldots$. Generalize this result.

Exercise 7.12.2 In base 10, show $(3)^* \cdot (.3333\ldots) = 1^*$.

Exercise 7.12.3 In base n, show:

1. $0^* \cdot A = 0^*$;

2. $1^* \cdot A = A$.

Exercise 7.12.4 In base 10, show $2^* \cdot A = A + A$.

We have said that finite and terminating decimals behave alike. This extends to multipation too.

Exercise 7.12.5 In base 10, as finite decimals, $.14 \times .28 = .0392$. Show that as infinite decimals $.140000\ldots \times .280000\ldots = .03920000\ldots$.

Exercise 7.12.6 Show that in base n, if d_1 and d_2 are finite decimals, $(d_1)^* \cdot (d_2)^* = (d_1 \cdot d_2)^*$.

Then, any principle of finite decimals involving only order, addition and multiplication can be transferred to a principle of terminating decimals. The following exercise illustrates this.

Exercise 7.12.7 Show the distributive law is true for base n terminating decimals. That is, let a, b and c be base n finite decimals and show $a^* \cdot (b^* + c^*) = a^* \cdot b^* + a^* \cdot c^*$.

7.13   Basic properties of multiplication

In this section we establish that multiplication of infinite decimals is commutative, associative, is related to addition by a distributive law, and we see how multiplication and the ordering relation are connected. Most of the proofs are quite similar to the corresponding proofs for addition and are left as exercises.

Theorem 7.13.1 (Commutativity of multiplication)
In base n, let A and B be infinite decimals. $A \cdot B = B \cdot A$.

Theorem 7.13.2 *In base n, let A, B and C be infinite decimals. If $A \geq B$ then $A \cdot C \geq B \cdot C$.*

Next we need some preliminary results to lead up to a proof of associativity of multiplication.

Lemma 7.13.3 *In base n, suppose U, V and T are finite decimals. Suppose also that $U \cdot V > T$. Then there are finite decimals a and b such that $U > a$, $V > b$ and $a \cdot b > T$.*

Proof Since U, V and T are finite decimals, we have all the basic properties of rational numbers at our disposal. $U \cdot V > T \geq 0$, so neither U nor V can be 0, hence we can divide by them. Now $U \cdot V > T$ so $U > \frac{T}{V}$. Let a be a finite decimal between U and $\frac{T}{V}$ (Exercise 6.4.2). Then $U > a$, and $a > \frac{T}{V}$, so $a \cdot V > T$. Further, since $a \cdot V > T$, $V > \frac{T}{a}$ (why can't a be 0?). Let b be a finite decimal between V and $\frac{T}{a}$. Then $V > b$, and $b > \frac{T}{a}$, so $a \cdot b > T$. □

Lemma 7.13.4 (Key Lemma) *In base n, let A and B be infinite decimals and T be a terminating decimal. Suppose $A \cdot B > T$. Then there are terminating decimals a and b such that $A > a$, $B > b$, and $a \cdot b > T$.*

Proof We may suppose A and B are in standard form. Let

$$A = w_A.a_1a_2a_3a_4 \ldots$$

$$B = w_B.b_1b_2b_3b_4 \ldots$$

Now $A \cdot B$ is the limit of the sequence

$$L_0 = [w_A \cdot w_B]^*$$
$$L_1 = [(w_A.a_1) \cdot (w_B.b_1)]^*$$
$$L_2 = [(w_A.a_1a_2) \cdot (w_B.b_1b_2)]^*$$
etc.

and also, $A \cdot B$ is the least upper bound of this sequence. $A \cdot B > T$, so for some j, $L_j > T$. That is,

$$[(w_A.a_1a_2\ldots a_j) \cdot (w_B.b_1b_2\ldots b_j)]^* > T.$$

By Exercise 7.12.6 this says

$$(w_A.a_1a_2\ldots a_j)^* \cdot (w_B.b_1b_2\ldots b_j)^* > T.$$

Let $U = (w_A.a_1a_2\ldots a_j)^*$ and $V = (w_B.b_1b_2\ldots b_j)^*$. Then U and V are terminating decimals. $U \cdot V > T$, and clearly $A \geq U$ and $B \geq V$.

Now, by the lemma above, transferred to terminating decimals, there are terminating decimals a and b such that $U > a$, $V > b$ and $a \cdot b > T$. Since $A \geq U > a$ we have $A > a$. Similarly $B > b$. This concludes the proof. □

Theorem 7.13.5 (Associativity of Multiplication)
In base n, let A, B and C be infinite decimals; $A \cdot (B \cdot C) = (A \cdot B) \cdot C$.

Theorem 7.13.6 (Distributivity) *In base n let A, B and C be infinite decimals; $A \cdot (B + C) = A \cdot B + A \cdot C$.*

Proof If any of A, B or C is 0^* the result is immediate. Now suppose none is 0^*.

We can show this result by showing, for each terminating decimal T, $A \cdot (B + C) > T$ if and only if $A \cdot B + A \cdot C > T$. Well, let T be a terminating decimal, and suppose first that $A \cdot B + A \cdot C > T$. By the Key Lemma of Section 11 there are terminating decimals d and e such that $A \cdot B > d$, $A \cdot C > e$ and $d + e > T$. Further, $A \cdot B > d$, so by the Key Lemma in this section there are terminating decimals a_1 and b such that $A > a_1$, $B > b$ and $a_1 \cdot b > d$. Similarly, since $A \cdot C > e$, there are terminating decimals a_2 and c such that $A > a_2$, $C > c$ and $a_2 \cdot c > e$. Let a be the larger of a_1 and a_2. Then $A > a$, $a \geq a_1$, $a \geq a_2$.

Now, $A \geq a$, $B \geq b$ and $C \geq c$, so $B + C \geq b + c$ and $A \cdot (B + C) \geq a \cdot (b + c)$. But the distributive law holds for *terminating decimals* so $a \cdot (b + c) = a \cdot b + a \cdot c$. But also, $a \geq a_1$ and $a \geq a_2$, so $a \cdot b \geq a_1 \cdot b$ and $a \cdot c \geq a_2 \cdot c$ so $a \cdot b + a \cdot c \geq a_1 \cdot b + a_2 \cdot c \geq d + e > T$. Thus $A \cdot (B + C) \geq a \cdot (b + c) = a \cdot b + a \cdot c > T$.   □

Theorem 7.13.7 *In base n, let A and B be infinite decimals. Let $\mathbf{C}_A$ be any collection of infinite decimals having A as least upper bound, and let $\mathbf{C}_B$ be any collection having B as least upper bound. Let $\mathbf{C}$ consist of all products of terms from $\mathbf{C}_A$ with terms from $\mathbf{C}_B$. That is, $\mathbf{C} = \{x \cdot y \mid x \in \mathbf{C}_A \text{ and } y \in \mathbf{C}_B\}$. Then $\mathbf{C}$ has $A \cdot B$ as least upper bound.*

Corollary 7.13.8 *In base n, let A_1, A_2, A_3, $A_4, \ldots$ be any bounded, non-decreasing sequence having A as limit, and let B_1, B_2, B_3, $B_4, \ldots$ similarly have B as limit. Then the sequence $A_1 \cdot B_1$, $A_2 \cdot B_2$, $A_3 \cdot B_3, \ldots$ has $A \cdot B$ as limit.*

When we multiply infinite decimals the definition requires that they be in standard form, but in fact this is not necessary, though it did serve to simplify some proofs. The demonstration of this is left to you in a series of exercises.

Exercises

Exercise 7.13.1 Prove Theorem 7.13.1.

Exercise 7.13.2 Use the corresponding theorem in Section 11 as a guide and give a proof of Theorem 7.13.2.

Exercise 7.13.3 Let A, B, C and D be base n infinite decimals. Suppose $A \geq C$ and $B \geq D$, show $A \cdot B \geq C \cdot D$.

Exercise 7.13.4 Use the proof of the associativity of addition in Section 11 as a guide and give a proof of Theorem 7.13.5.

Exercise 7.13.5 Complete the proof of Theorem 7.13.6 by showing: if $A \cdot (B + C) > T$ where T is a terminating decimal, then $A \cdot B + A \cdot C > T$.

Exercise 7.13.6 Prove Theorem 7.13.7.

Exercise 7.13.7 Prove Corollary 7.13.8.

Exercise 7.13.8 Carry out the following base 2 multiplication twice, once using standard forms, once using non-standard forms:

$$1.110000\ldots \times 11.100000\ldots.$$

Exercise 7.13.9 Prove that restrictions to standard forms in the definition of multiplication is not necessary (see the corresponding discussion of addition at the end of Section 11.

7.14 Subtraction, an introduction

It is easy to say how subtraction *should* behave. If $A \geq B$, $A - B$ should be that quantity C such that $B + C = A$. The difficulty in simply making this the definition of subtraction is that it is not easy to establish that whenever $A \geq B$, there really is exactly one C such that $B + C = A$.

One way out of this difficulty is to define subtraction by totally different means, using intuition and analogy as guides, then develop its properties to the point where we can show it meets the condition given at the beginning of this section. Now there are several different, but equivalent ways this can be done. Some are more convenient than others for the purposes of a simple development of the theory. As it turns out, the way that is most analogous to the way we defined addition and multiplication leads to unexpected difficulties; suppose, for example, we had the base 10 problem.

$$
\begin{array}{r}
1.729365\ldots \\
- \quad 1.682737\ldots \\
\hline
\end{array}
$$

and suppose we try subtracting better and better finite decimal approximations and see if the results 'go anywhere.' Well, as finite decimals,

$$
\begin{array}{rcl}
1. - 1. & = & 0. \\
1.7 - 1.6 & = & 0.1 \\
1.72 - 1.68 & = & 0.04 \\
1.729 - 1.682 & = & 0.047 \\
1.7293 - 1.6827 & = & 0.0466 \\
1.72936 - 1.68273 & = & 0.04663 \\
1.729365 - 1.682737 & = & 0.046628 \\
\text{etc.} & & \text{etc.}
\end{array}
$$

Notice that is sequence isn't non-decreasing (nor, for that matter is it non-increasing!). Rather it oscillates. The second term is bigger than the first, the third term smaller than the second, then the fourth is bigger and so on. We can not assign a limit to this sequence by any means we have established so far. Our approach to subtraction must be along different lines.

7.15 Subtraction with terminating decimals

As we have seen, subtraction with infinite decimals is something of a problem, but oddly enough if at least one of the decimals involved is terminating the situation becomes intuitively simple. Let us first consider a subtraction problem with a terminating decimal 'on the bottom.' Consider, in base 10, the problem:

$$
\begin{array}{r}
2.7163092816\ldots \\
-\quad .6319000000\ldots \\
\hline
\end{array}
$$

It is intuitively plausible to say: subtract as finite decimals:

$$
\begin{array}{r}
2.7163 \\
-\quad .6319 \\
\hline
2.0844
\end{array}
$$

and 'tack on' the rest of the upper decimal, namely $092816\ldots$, to get the infinite decimal $2.0844092816\ldots$. Thus we might write:

$$
\begin{array}{r}
2.7163092816\ldots \\
-\quad .6319000000\ldots \\
\hline
2.0844092816\ldots
\end{array}
$$

One is justified in this by showing the answer 'adds back',

$$
\begin{array}{r}
2.0844092816\ldots \\
+\quad .6319000000\ldots \\
\hline
2.7163092816\ldots
\end{array}
$$

In general, in base n, let A be an infinite decimal and T be a terminating decimal, $(A > T)$. We define $A - T$ as follows. Suppose the standard forms of A and T are:

$$A = w_A.a_1 a_2 a_3 \ldots a_n a_{n+1} \ldots$$

$$T = w_T.t_1 t_2 t_3 \ldots t_n 0000 \ldots$$

Compute as a finite decimal problem

$$
\begin{array}{r}
w_A.a_1 a_2 a_3 \ldots a_n \\
-\quad w_T.t_1 t_2 t_3 \ldots t_n \\
\hline
w_C.c_1 c_2 c_3 \ldots c_n
\end{array}
$$

Then we define $A - T$ to be the infinite decimal

$$C = w_C.c_1 c_2 c_3 \ldots c_n a_{n+1} a_{n+2} \ldots.$$

Note that, differently phrased, Exercise 7.15.2 shows $(A - T) + T = A$.

Now let us consider the problem of subtracting an infinite decimal *from* a terminating decimal. For example, consider the base 10 problem:

$$3.240000\ldots$$
$$-\quad 1.826391\ldots$$

Let us begin by replacing the terminating decimal on top by its nonstandard form, in effect doing all our borrowing at once.

$$3.239999\ldots$$
$$-\quad 1.826391\ldots$$

Again we can compute, as finite decimals,

$$3.23$$
$$-\quad 1.82$$

Now notice, in our original problem, after the second decimal place borrowing is never necessary. So we can compute column by column independently, and put down:

$$3.239999\ldots$$
$$-\quad 1.826391\ldots$$
$$\overline{1.413608\ldots}$$

Again, the procedure is justified if we show it 'adds back.' This is left to you as an Exercise.

In general, in base n, let A be an infinite decimal and T be a terminating decimal $(T > A)$. We define $T - A$ as follows. Let:

$$T = u_T.t_1t_2\ldots t_{k-1}t_k0000\ldots$$

$$A = w_A.a_1a_2\ldots a_{k-1}a_ka_{k+1}\ldots$$

(here we suppose t_k is the last non-zero decimal place of T.) Then T in non-standard form is:

$$T = u_T.t_1t_2\ldots t_{k-1}t_k'dddd\ldots$$

where d is the highest base n digit and t_k' is the digit just smaller than t_k. Now, compute as a finite decimal problem:

$$w_T.t_1t_2\ldots t_{k-1}t_k'$$
$$-\quad w_A.a_1a_2\ldots a_{k-1}a_k$$
$$\overline{w_C.c_1c_2\ldots c_{k-1}c_k}$$

Also let:

$$c_{k+1} = d - a_{k+1}$$

$$c_{k+2} = d - a_{k+2}$$

$$c_{k+3} = d - a_{k+3}$$

etc.

and set

$$C = w_C.c_1c_2\ldots c_{k-1}c_kc_{k+1}c_{k+2}\cdots$$

We define $T - A$ to be C.

Differently phrased, Exercise 7.15.4 shows $(T - A) + A = T$.

We have now defined $A - B$ $(A > B)$ in the case that one of A or B is a terminating decimal. In fact, if both are terminating decimals, $A - B$ is covered by both discussions above, so it is necessary to show that each method gives the same answer in this case. Exercise 7.15.5 illustrates this.

A general proof that our two methods of subtraction agree when both decimals involved are terminating is quite easy. First, we know that in the system of rational numbers (hence in finite decimals) if $a + c = b$ and $a + c' = b$ then $c = c'$ (cancellation law for addition of rationals). Then a similar result holds for terminating decimals. Now suppose T_1 and T_2 are terminating decimals with $T_1 > T_2$. Clearly either method of subtracting $T_1 - T_2$ will give a terminating decimal for a result. Say the first method discussed in this section gives us $T_1 - T_2 = C$, and say the second method produces $T_1 - T_2 = C'$. By Exercise 7.15.2, $C + T_2 = T_1$, and by Exercise 7.15.4, $C' + T_2 = T_1$. But then, since we have the cancellation law for terminating decimals, we get $C = C'$.

Exercises

Exercise 7.15.1 Use the definition of addition we gave for infinite decimals, and verify the correctness of the addition:
$2.0844092816\ldots + .6319000000 = 2.7163092816\ldots$.

Exercise 7.15.2 Let $A = w_A.a_1a_2a_3\ldots a_na_{n+1}\cdots$,
$T = w_T.t_1t_2t_3\ldots t_n0000\ldots$ and $C = w_C.c_1c_2c_3\ldots c_na_{n+1}a_{n+2}\cdots$ be as above, so that $A - T = C$ by our definition. Prove that $C + T = A$.

Exercise 7.15.3 Show that $1.413608\ldots + 1.826391\ldots = 3.239999\ldots$, which equals $3.240000\ldots$.

Exercise 7.15.4 Let $A = w_A.a_1a_2\ldots a_{k-1}a_ka_{k+1}\cdots$,
$T = w_T.t_1t_2\ldots t_{k-1}t_k0000\ldots$ and $C = w_C.c_1c_2\ldots c_{k-1}c_kc_{k+1}c_{k+2}\cdots$, so that $T - A = C$ by our definition. Prove that $C + A = T$.

Exercise 7.15.5 In base 10, carry out the following subtraction twice, using both methods discussed in this section:$18.360000\ldots - 4.823000\ldots$.

7.16 Subtraction in general

Suppose A and B are base n infinite decimals with $A \geq B$. We want to say what $A - B$ is to mean even if neither A nor B is terminating. Our definition is this. First, if $A = B$, we define $A - B$ to be 0. Next, if $A > B$, there is a terminating decimal T between them (Theorem 7.7.5): $A > T > B$. In the previous section, both $A - T$ and $T - B$ were defined. We take $A - B$ to be $(A - T) + (T - B)$.

For example, using this definition let us compute in base 10:

$$
\begin{aligned}
4.2183692\ldots &= A \\
-\quad 4.1892136\ldots &= B
\end{aligned}
$$

In Section 7 our proof that there was a terminating decimal between A and B actually showed how to produce one. Following the method of that proof, we get $T = 4.190000\ldots$ Now, we compute $A - T$ as defined in Section 15:

$$
\begin{aligned}
4.2183692\ldots & \\
-\quad 4.1900000\ldots & \\
\hline
0.0283692\ldots &
\end{aligned}
$$

and we compute $T - B$:

$$
\begin{aligned}
4.1899999\ldots & \\
-\quad 4.1892136\ldots & \\
\hline
0.0007863\ldots &
\end{aligned}
$$

Then by our definition, $A - B = (A - T) + (T - B)$ or

$$
\begin{aligned}
0.0283692\ldots & \\
+\quad 0.0007863\ldots & \\
\hline
0.0291555\ldots &
\end{aligned}
$$

If $A > B$, we have defined $A - B$ to be $(A - T) + (T - B)$ where T is a terminating decimal between A and B. *But* we are not really free to use this definition until we clear up some difficulties. Would a different choice of terminating decimal between A and B give us a different result? Is it really the case that $A - B$ behaves like subtraction should, so that $(A - B) + B = A$? Is it the case that any other way of defining subtraction which makes $(A - B) + B = A$ would give the same result our way gives? We proceed to answer these questions, but until we do, we will not use the notation $A - B$ except in the special cases which were covered in Section 15.

Recall that in Section 15, subtraction was properly defined provided one of the decimals involved was terminating. Indeed, if $A > T > B$ where T is terminating, the subtraction operations of Section 15 satisfy:

$$
(A - T) + T = A
$$

$$
(T - B) + B = T
$$

Lemma 7.16.1 *In base n, let A be an infinite decimal and T be a terminating decimal. If $A > T$ then $A - T > 0^*$.*

Proof If A is a terminating decimal too, the result is immediate, since it is true for finite decimals.

If A is not a terminating decimal, say its standard form is

$$w_A.a_1 a_2 \ldots a_n a_{n+1} \ldots$$

and T is

$$w_T.t_1 t_2 \ldots t_n 0000 \ldots.$$

Now, as defined in Section 15, to compute $A - T$ we carry out the finite decimal problem:

$$
\begin{array}{r}
w_A.a_1 a_2 \ldots a_n \\
-\quad w_T.t_1 t_2 \ldots t_n \\
\hline
w_C.c_1 c_2 \ldots c_n
\end{array}
$$

Then $A - T$ is defined to be

$$w_C.c_1 c_2 \ldots c_n a_{n+1} a_{n+2} \ldots$$

But not all of $a_{n+1}, a_{n+2}, \ldots$ can be 0's since A is not terminating. Hence $A - T > 0^*$. □

Lemma 7.16.2 *In base n, if $X > 0^*$ then $Y + X > Y$.*

Theorem 7.16.3 (Cancellation law for addition)
In base n, if $C + A = C + B$ then $A = B$.

Proof Suppose $C + A = C + B$. If we didn't have $A = B$ we would have $A > B$ or $B > A$, say the first for the sake of argument. Let T be a terminating decimal between A and B, thus $A > T > B$. Then in Section 15, $A - T$ and $T - B$ have been defined. Now $C + B = C + A$ so $C + B + (T - B) = C + A + (T - B)$. But $B + (T - B) = T$, so we have $C + T = C + A + (T - B) \geq C + A + 0^* = C + A$. From this using Theorem 7.11.2, $C + T + (A - T) \geq C + A + (A - T)$. But $T + (A - T) = A$, so we have $C + A \geq C + A + (A - T)$. But now, by Lemma 7.16.1, $A - T > 0^*$, so by Lemma 7.16.2, $(C + A) + (A - T) > (C + A)$. Thus $C + A > C + A$ and this is not possible. A similar contradiction arises if $B > A$ so the proof is complete. □

The cancellation law is an important result. One consequence is that subtraction, if it can be done at all, can only give one answer. That is,

Corollary 7.16.4 *In base n, if there is some C such that $B + C = A$, there is only one.*

Next we establish when subtraction can be done, and show the way we defined it is proper.

Theorem 7.16.5 *In base* n, *let* $A > B$. *Let* T *be some terminating decimal between* A *and* B, *and let* $C = (A - T) + (T - B)$. *Then:*

1. $B + C = A$

2. *If* T' *is any other terminating decimal between* A *and* B, $(A - T') + (T' - B) = C$.

Proof For 1)

$$
\begin{aligned}
B + C &= B + [(A - T) + (T - B)] \\
&= B + [(T - B) + (A - T)] \\
&= [B + (T - B)] + (A - T) \\
&= T + (A - T) \\
&= A.
\end{aligned}
$$

And for 2), we merely observe that our proof of part 1) will work for any terminating decimal between A and B, and so

$$ B + [(A - T') + (T' - B)] = A. $$

But $B + C = A$, so

$$ B + C = B + [(A - T') + (T' - B)] $$

and by the cancellation law,

$$ C = (A - T') + (T' - B). $$

$\square$

Thus our definition of subtraction is a good one. If $A \geq B$, $A - B$ is the one and only thing we can add to B and get A. That is,

1. $B + (A - B) = A$

2. if $B + X = A$ then $X = A - B$.

We conclude this section with a proof that finite and terminating decimals behave alike with respect to subtraction as well.

Theorem 7.16.6 *Let* d_1 *and* d_2 *be base* n *finite decimals with* $d_1 \geq d_2$. *Then* $(d_1 - d_2)^* = d_1^* - d_2^*$.

Proof We have seen that $d_1^* - d_2^*$ is the only thing we can add to d_2^* to get d_1^*. But suppose we try adding $(d_1 - d_2)^*$ to d_2^* and see what happens.

$$ d_2^* + (d_1 - d_2)^* = [d_2 + (d_1 - d_2)]^* $$

since finite and terminating decimals behave alike with respect to addition. But, as *finite* decimals, $d_2 + (d_1 - d_2) = d_1$ so $[d_2 + (d_1 - d_2)]^* = d_1^*$. Thus $d_2^* + (d_1 - d_2)^* = d_1^*$ so $(d_1 - d_2)^* = d_1^* - d_2^*$. $\square$

We now have several ways available for subtracting terminating decimals. There are the techniques of Section 15, and the general method of this section. In fact, they all give the same results. We leave it to you to verify this.

Exercises

Exercise 7.16.1 In base 2, compute the following:

$$11.0110110010110\ldots$$
$$- \quad 1.0001101001101\ldots$$

Exercise 7.16.2 Prove that if A and B are base n infinite decimals with $A > B$, there are infinitely many terminating decimals between them.

Exercise 7.16.3 Prove Lemma 7.16.2.

Exercise 7.16.4 Prove Corollary 7.16.4.

Exercise 7.16.5 If $A = B$, we defined $A - B$ to be 0^*. Show 0^* is the only thing satisfying $B + C = A$.

Exercise 7.16.6 In base 10 calculate:

$$4.30000\ldots$$
$$- \quad 3.10000\ldots$$

Use the method of this section; that is, find a terminating decimal between them, etc. Also subtract by one other method.

Exercise 7.16.7 Prove that all the definitions of subtraction, applied to terminating decimals, give the same results.

7.17   More results on the ordering of infinite decimals

We give another characterization of the ordering relation for infinite decimals, and we sharpen some earlier results.

Theorem 7.17.1 *In base* n,

1. *$A \geq B$ if and only if $A = B + C$ for some C,*

2. *$A > B$ if and only if $A = B + C$ for some $C \neq 0^*$.*

Proof Suppose first that $A \geq B$. Then $A - B$ is defined. Let $C = A - B$. Then $B + C = A$.

Suppose next that $A = B + C$ for some C. Well, $C \geq 0^*$, so $A = B + C \geq B + 0^* = B$ so $A \geq B$. Thus part 1) is established.

Now suppose $A > B$. Again, $A - B$ is defined. Let T be a terminating decimal between A and B, $A > T > B$. Then $A - B = (A - T) + (T - B)$ by definition. By Lemma 7.16.1, $A - T > 0^*$, so $A - B = (A - T) + (T - B) \geq (A - T) + 0^* = A - T > 0^*$, that is, $A - B > 0^*$. Now take $C = A - B$. Then $B + C = A$ and $C \neq 0^*$. This is half of part 2).

Finally suppose $A = B + C$ for some $C \neq 0^*$. By Lemma 7.16.2, since $C > 0^*$, $B + C > B$. Thus $A > B$. This completes the proof. $\square$

We showed in Section 11 that if $A \geq B$ then $A + C \geq B + C$. Now we are in a position to show that $\geq$ can be replaced by $>$, which is a sharper result. Also, the converse holds, which is quite useful.

Theorem 7.17.2 *In base n, $A > B$ if and only if $A + C > B + C$.*

Proof Suppose $A > B$. Then certainly $A \geq B$, so $A + C \geq B + C$. If we did not have $A + C > B + C$, we would have $A + C = B + C$ and by the Cancellation Law of Section 4 we would have $A = B$, a contradiction. Thus $A + C > B + C$.

Next suppose $A + C > B - C$. If we did not have $A > B$, we would have $B \geq A$. But then, $B + C \geq A + C$, contradicting that $A + C > B + C$. Thus $A > B$. □

Now we can get a similar result to Theorem 7.17.2 for multiplication.

Theorem 7.17.3 *In base n, let A, B and K be infinite decimals with $K \neq 0^*$. Then:*

1. $A = B$ if and only if $K \cdot A = K \cdot B$,

2. $A > B$ if and only if $K \cdot A > K \cdot B$.

Proof We show part 2). Suppose first that $A > B$. By Theorem 7.17.1, $A = B + C$ where $C \neq 0^*$. Then $K \cdot A = K \cdot (B + C) = K \cdot B + K \cdot C$. Now by Exercise 7.17.2, $K \cdot C \neq 0^*$, so by Theorem 7.17.1 again, $K \cdot A > K \cdot B$.

Conversely, if $K \cdot A > K \cdot B$ we must have $A > B$, for otherwise $A = B$ or $B > A$. If $A = B$ then $K \cdot A = K \cdot B$, a contradiction. If $B > A$, by what we just showed, $K \cdot B > K \cdot A$, again a contradiction. Thus $A > B$. □

We have shown that the same thing can be added to both sides of an inequality. In fact, the same thing can be subtracted from both sides as well.

Theorem 7.17.4 *In base n, if $A \geq C$ and $B \geq C$, then $A > B$ if and only if $A - C > B - C$.*

Proof Suppose $A > B$. If we didn't have $A - C > B - C$, then we would have $B - C \geq A - C$. But then $(B - C) + C \geq (A - C) + C$, or $B \geq A$, a contradiction. Thus $A - C > B - C$. This is half of the theorem. The other half is left as an exercise. □

On the other hand we have:

Theorem 7.17.5 *In base n, if $C \geq A$ and $C \geq B$ then, $A > B$ if and only if $C - A < C - B$.*

Proof Suppose $A > B$. If we didn't have $C - A < C - B$ we would have $C - A \geq C - B$. But then, $(C - A) + A \geq (C - B) + A > (C - B) + B$. But $(C - A) + A = C$ and $(C - B) - B = C$ so we have $C > C$, a contradiction. □

Exercises

Exercise 7.17.1 Prove, in base n, $A \geq B$ if and only if $A + C \geq B + C$.

Exercise 7.17.2 In base n, prove that if $A \neq 0^*$ and $B \neq 0^*$ then $A \cdot B \neq 0^*$.

Exercise 7.17.3 Prove part 1) of Theorem 7.17.3.

Exercise 7.17.4 Prove or disprove: the restriction, $K \neq 0^*$ stated in Theorem 7.17.3 can be dropped.

Exercise 7.17.5 Prove, in base n, if $A > B$ and $C > D$ then $A \cdot C > B \cdot D$.

Exercise 7.17.6 Prove, in base n, that each of the following implies the next.

$$A > 1$$
$$A \cdot A > 1$$
$$A \cdot A \cdot A > 1$$
$$A \cdot A \cdot A \cdot A > 1$$
etc.

Exercise 7.17.7 Prove, in base n, that each of the following implies the next.

$$A < 1$$
$$A \cdot A < 1$$
$$A \cdot A \cdot A < 1$$
$$A \cdot A \cdot A \cdot A < 1$$
etc.

Exercise 7.17.8 Prove, in base n, that all of the following are equivalent.

$$A > 1$$
$$A \cdot A > 1$$
$$A \cdot A \cdot A > 1$$
$$A \cdot A \cdot A \cdot A > 1$$
etc.

Exercise 7.17.9 Complete the proof of Theorem 7.17.4 by showing, if $A - C > B - C$ then $A > B$.

Exercise 7.17.10 Complete the proof of Theorem 7.17.5 by showing: if $C - A < C - B$ then $A > B$.

Exercise 7.17.11 In base n, suppose $A \geq C$ and $B \geq D$ and show: if $A > B$ and $C < D$ then $A - C > B - D$.

7.18 Basic properties of subtraction

The properties stated for subtraction here can be proved by the same methods that worked for the whole number system, see Chapter Three Section 7. Consequently they are left as exercises.

By definition, $A - B = (A - T) + (T - B)$ where T is a terminating decimal between A and B. In fact, the restriction to terminating can be removed.

Theorem 7.18.1 *In base n, if $A \geq C \geq B$ then $A - B = (A - C) + (C - B)$.*

Theorem 7.18.2 (Distributive law for subtraction)
In base n, if $B \geq C$, then $A \cdot (B - C) = A \cdot B - A \cdot C$.

Exercises

Exercise 7.18.1 Prove Theorem 7.18.1.

Exercise 7.18.2 Prove Theorem 7.18.2.

Exercise 7.18.3 In base n, if $B \geq C$ then $(A + B) - C = A + (B - C)$.

Exercise 7.18.4 In base n, if $A \geq B + C$ then $A - (B + C) = (A - B) - C = (A - C) - B$.

Exercise 7.18.5 In base n, if $A \geq B$ then $A - B = (A + C) - (B + C)$.

Exercise 7.18.6 In base n, if $A \geq B \geq C$ then $A - (B - C) = (A - B) + C = (A + C) - B$.

Exercise 7.18.7 In base n, if $A \geq B$ and $C \geq D$ then $(A - B) \cdot (C - D) = (A \cdot C + B \cdot D) - (A \cdot D + B \cdot C)$.

Hint: It will be easiest to establish this using the properties of subtraction above, rather than working directly.

7.19 Division

We show that division, except by 0^*, always makes sense. That is, if A and B are base n infinite decimals with $B \neq 0^*$, then

1. there is an infinite decimal C such that $B \cdot C = A$,

2. there is only one.

Then we can define $A \div B$ to be that infinite decimal C such that $B \cdot C = A$. To show 1) we use the Least Upper Bound Theorem 7.9.7. To show 2) we use the cancellation law for multiplication, Theorem 7.17.3. We begin with the easy part, item 2.

Theorem 7.19.1 *In base n, let A and B be infinite decimals with $B \neq 0^*$. If there is an infinite decimal C such that $B \cdot C = A$, there is only one.*

Proof Suppose $B \cdot C = A$ and also $B \cdot C' = A$. Then $B \cdot C = B \cdot C'$ and by the cancellation law for multiplication, $C = C'$. □

For item 1 we first need a simple lemma whose proof says, multiplying by $(10)^*$ often enough makes a product as big as we like, and multiplying by $(0.1)^*$ often enough can make it as small as we like.

Lemma 7.19.2 *In base n, let B and D be infinite decimals, neither 0^*.*

1. *There is an infinite decimal K, not 0^*, such that $D > B \cdot K$.*

2. *There is an infinite decimal K, such that $B \cdot K > D$.*

Proof We show part 1 only. To simplify the language we suppose both B and D have 0 whole number parts. It will be clear how to modify things if that is not the case. So, say B and D in standard form look like

$$B = 0.b_1 b_2 b_3 b_4 \ldots$$

$$D = 0.d_1 d_2 d_3 d_4 \ldots$$

Now $D \neq 0^*$ so some decimal place of D is not 0, say the first non-zero place of D is the kth. Then D looks like:

$$D = 0.00 \ldots 0 d_k d_{k+1} \ldots$$

where $d_k \neq 0$. It is easy to show that multiplying an infinite decimal by $(0.1)^*$ moves the decimal point to the left one place. So, take K to be $(.1)^* \cdot (.1)^* \cdot \ldots \cdot (.1)^*$ (where we have written k terms in the product). Then multiplying by K moves the decimal point k places to the left. So b_1, being the first decimal place of B, becomes the $k + 1^{st}$ of $B \cdot K$. That is,

$$B \cdot K = 0.00 \ldots 0 b_1 b_2 b_3 b_4 \ldots$$

where there are k 0's after the decimal point. Now, since $B \cdot K$ has 0 in its kth decimal place while D has $d_k \neq 0$, $D > B \cdot K$. □

Theorem 7.19.3 *In base n, let A and B be infinite decimals with $B \neq 0^*$. There is an infinite decimal C such that $B \cdot C = A$.*

Proof We form a collection $\mathbf{C}$ of base n infinite decimals as follows: put X in $\mathbf{C}$ just when $A \geq B \cdot X$. That is, $\mathbf{C} = \{X \mid A \geq B \cdot X\}$.

First, since $A \geq B \cdot 0^* = 0^*$, $\mathbf{C}$ has something in it, namely 0^*.

Next, by part 2) of the lemma above, there is an infinite decimal K such that $B \cdot K > A$. We claim K is an upper bound for $\mathbf{C}$. That is, if X belongs to **C**, then $X < K$. For, if X is in **C**, then $B \cdot X \leq A$, but $A < B \cdot K$, so we have $B \cdot X < B \cdot K$, and by Theorem 7.17.3, $X < K$.

Thus **C** is a non-empty collection of infinite decimals which has an upper bound. Then **C** has a least upper bound, call it C. We claim C is what we wanted, that is, $B \cdot C = A$. We show this by showing both of $A > B \cdot C$ and $B \cdot C > A$ lead to contradictions.

Suppose we had $A > B \cdot C$. We show C can be increased a little to C' and still have $A > B \cdot C'$, which means C is not an upper bound for **C**.

Well, if $A > B \cdot C$, then $A - B \cdot C$ is defined and is not 0^*. Call it D. By part 1) of the lemma above, there is an infinite decimal K, not 0^*, such that $D > B \cdot K$. Let $C' = C + K$. Then:

$$
\begin{aligned}
A &= B \cdot C + (A - B \cdot C) \\
&= B \cdot C + D \\
&> B \cdot C + B \cdot K \\
&= B \cdot (C + K) \\
&= B \cdot C'
\end{aligned}
$$

so $A > B \cdot C'$ which means C' is in **C**. But $C' > C$ and C was an upper bound, a contradiction.

Suppose we had $B \cdot C > A$. We show C can be decreased a little to c and still have $B \cdot c > A$, which means C is not the *least* upper bound. This can be done in a manner similar to the case just considered, but here we have an earlier result at our disposal which we use for variety. Well, suppose $B \cdot C > A$. There is a terminating decimal T between then, $B \cdot C > T > A$. By Lemma 7.13.4, there are terminating decimals b and c with $B > b$, $C > c$ and $b \cdot c > T$. Then since $B > b$, $B \cdot c > b \cdot c > T > A$ so $B \cdot c > A$, and hence c is an upper bound for **C** (why?) but $c < C$ which is the least upper bound, again a contradiction.

This concludes the proof. □

Definition 7.19.4 In base n, let A and B be infinite decimals with $B \neq 0^*$. By $A \div B$ we mean that infinite decimal C such that $B \cdot C = A$.

Notice that, by definition, $B \cdot (A \div B) = A$, and $B \cdot X = A$ implies $X = A \div B$.

Exercises

Exercise 7.19.1 Prove part 2) of Lemma 7.19.2 by a similar argument, involving multiplying by $(10)^*$ instead of by $(0.1)^*$.

Exercise 7.19.2 According to our definition, $A \div B$ makes sense if $B \neq 0^*$. Show this restriction is necessary. Consider two cases, $A = 0^*$ and $A \neq 0^*$.

Exercise 7.19.3 Prove, in base 10, $(6.0000\ldots) \div (3.0000\ldots) = 2.0000\ldots$.

Exercise 7.19.4 Prove, in base 10, $(1.0000\ldots) \div (3.0000\ldots) = 0.3333\ldots$.

Exercise 7.19.5 Prove, in base 2,

$$(1.0000\ldots) \div (11.0000\ldots) = 0.010101\ldots$$

.

7.20   Basic properties of division

Most of what we need to know about division follows quickly from what we already know about multiplication. We begin by presenting a notation for division that has become the conventional one. Its similarity to fraction notation in the rational number system should arouse suspicions which, in fact, will be justified shortly.

Definition 7.20.1 If A and B are base n infinite decimals with $B \neq 0^*$ we write $\frac{A}{B}$ for $A \div B$.

Note that, by definition of division, $B \cdot \frac{A}{B} = A$, and $B \cdot X = A$ implies $X = \frac{A}{B}$.

We now officially adopt the convention that whenever we write $\frac{A}{B}$ we assume $B \neq 0^*$. This cuts down on the wording of theorems and proofs.

Theorem 7.20.2 *In base* n, $\frac{A}{B} = A \cdot \frac{(1)^*}{B}$.

Proof We know $B \cdot \frac{A}{B} = A$. But also

$$
\begin{aligned}
B \cdot \left[A \cdot \tfrac{(1)^*}{B} \right] &= A \cdot \left[B \cdot \tfrac{(1)^*}{B} \right] \\
&= A \cdot (1)^* \\
&= A.
\end{aligned}
$$

Hence

$$B \cdot \left[A \cdot \tfrac{(1)^*}{B} \right] = B \cdot \tfrac{A}{B}$$

so by the cancellation law for multiplication,

$$\frac{A}{B} = A \cdot \frac{(1)^*}{B}.$$

$\square$

This suggests one more piece of notation.

Definition 7.20.3 We write B^{-1} for $\frac{(1)^*}{B}$. B^{-1} is read 'B inverse'.

Note that the theorem above, using this notation, reads $\frac{A}{B} = A \cdot B^{-1}$. This means a division can be replaced by a multiplication.

Theorem 7.20.4 (cross multiplication) *In base n, $\frac{A}{B} = \frac{C}{D}$ if and only if $A \cdot D = B \cdot C$.*

Proof Suppose first that $\frac{A}{B} = \frac{C}{D}$, that is, $A \cdot B^{-1} = C \cdot D^{-1}$. Then:

$$
\begin{aligned}
A \cdot B^{-1} \cdot B &= C \cdot D^{-1} \cdot B \\
A \cdot 1^* &= C \cdot D^{-1} \cdot B \\
A &= B \cdot C \cdot D^{-1}
\end{aligned}
$$

also

$$
\begin{aligned}
A \cdot D &= B \cdot C \cdot D^{-1} \\
A \cdot D &= B \cdot C \cdot 1^* \\
A \cdot D &= B \cdot C.
\end{aligned}
$$

This is half the proof, the other half is similar, starting with $A \cdot D = B \cdot C$ and multiplying by B^{-1} and D^{-1}. □

The proof of this theorem is typical. We use inverse notation to replace divisions by multiplications, then things are easy. We work out one more theorem in detail and leave the rest as exercises.

Theorem 7.20.5 *In base n, $\frac{A}{C} + \frac{B}{C} = \frac{A+B}{C}$.*

Proof The theorem, rewritten, states $A \cdot C^{-1} + B \cdot C^{-1} = (A + B) \cdot C^{-1}$ and this is true by the distributive law. □

Exercises

Exercise 7.20.1 Show $B \cdot B^{-1} = 1^*$.

Exercise 7.20.2 Show $\frac{A}{B} = \frac{A \cdot C}{B \cdot C}$.

Exercise 7.20.3 Show $\frac{A}{B} + \frac{C}{D} = \frac{A \cdot D + B \cdot C}{B \cdot D}$.

Exercise 7.20.4 Show $(B^{-1})^{-1} = B$.

Exercise 7.20.5 Show $\frac{A}{C} > \frac{B}{C}$ if and only if $A > B$.

Exercise 7.20.6 Show $\frac{A}{B} > \frac{C}{D}$ if and only if $A \cdot D > B \cdot C$.

Exercise 7.20.7 Show if $A > B$ then $A > \frac{A+B}{2^*} > B$.

Exercise 7.20.8 Show, if $A > C$, then $\frac{A}{C} - \frac{B}{C} = \frac{A-B}{C}$.

Exercise 7.20.9 Show, if $\frac{A}{B} \geq \frac{C}{D}$ then $\frac{A}{B} - \frac{C}{D} = \frac{A \cdot D - B \cdot C}{B \cdot D}$.

Exercise 7.20.10 Show $\frac{A}{B} \cdot \frac{C}{D} = \frac{A \cdot C}{B \cdot D}$.

Exercise 7.20.11 Show $\frac{A}{B} = 0^*$ if and only if $A = 0^*$.

Exercise 7.20.12 Show, if $\frac{A}{B} \neq 0^*$, then $\frac{A}{B} \cdot \frac{B}{A} = 1^*$.

Exercise 7.20.13 Show, if $\frac{A}{B} \neq 0^*$, then $[\frac{A}{B}]^{-1} = \frac{B}{A}$.

Exercise 7.20.14 Show, if $\frac{C}{D} \neq 0^*$, then $\frac{A}{B} \div \frac{C}{D} = \frac{A}{B} \cdot \frac{D}{C}$.

7.21   Rational numbers and infinite decimals

The rational number system is an enormously useful one, and infinite decimals would be less than satisfactory if, in creating them, we lost the rationals. But in fact the system of base n infinite decimals contains a very natural copy of the entire system of rational numbers, including those that are not base n finite decimals. This is easy to show now that division has been discussed.

First we need that the system of base n infinite decimals contains a copy of the whole number system. This is immediate since whole numbers are available as base n finite decimals. Recall that if t is a finite decimal, t^* is the terminating (infinite) decimal associated with it. Now we claim that if a and b are whole numbers, the infinite decimals a^* and b^* behave like a and b. For instance, in base 10, $2^* + 3^* = 5^*$. More generally, we have already seen that, in any base,

$$(a^*) + (b^*) = (a + b)^*$$

$$(a^*) \cdot (b^*) = (a \cdot b)^*$$

$$(a^*) > (b^*) \text{ if and only if } a > b$$

$$\text{if } a \geq b \text{ then } (a^*) - (b^*) = (a - b)^*.$$

These items are sufficient to verify our claim.

Now we want to find base n infinite decimal counterparts for all the rational numbers. Well, choose a rational number and let $\frac{a}{b}$ be a base n fraction naming it. We claim the base n infinite decimal $\frac{a^*}{b^*}$ will 'behave like' the rational number $\frac{a}{b}$.

For example, in base 10, the infinite decimal $\frac{1^*}{3^*}$ should behave like the rational number $\frac{1}{3}$. We know from Exercise 7.19.4 that $\frac{1^*}{3^*} = .3333\ldots$, so it is this infinite decimal we are asserting will act like $\frac{1}{3}$.

There is a difficulty here though. Rational numbers have many fraction names; could a different choice of fraction produce a different infinite decimal? For instance, in base 10, $\frac{1}{3} = \frac{2}{6}$. We know $\frac{1^*}{3^*} = .3333\ldots$, is it the case that $\frac{2^*}{6^*} = .3333\ldots$? The following theorem shows there is never an ambiguity here.

Theorem 7.21.1 *In base n, let $\frac{a}{b}$ and $\frac{c}{d}$ be fractions. $\frac{a}{b}$ and $\frac{c}{d}$ name the same rational number if and only if $\frac{a^*}{b^*}$ and $\frac{c^*}{d^*}$ are the same infinite decimal.*

Proof $\frac{a}{b}$ and $\frac{c}{d}$ name the same rational number if and only if $a \cdot d = b \cdot c$. But this is equivalent to $(a \cdot d)^* = (b \cdot c)^*$ which in turn is equivalent to $(a^*) \cdot (d^*) = (b^*) \cdot (c^*)$. Finally by Theorem 7.20.4 this is equivalent to $\frac{a^*}{b^*} = \frac{c^*}{d^*}$. □

So our association of infinite decimals with rational numbers makes sense. Next we need to show the infinite decimal $\frac{a^*}{b^*}$ and the rational number $\frac{a}{b}$ behave alike. For example, as base 10 fractions,

$$\frac{1}{3} + \frac{1}{3} + \frac{1}{3} = \frac{1}{1}$$

so in the system of base 10 infinite decimals, we should have

$$\frac{1^*}{3^*} + \frac{1^*}{3^*} + \frac{1^*}{3^*} = \frac{1^*}{1^*}$$

We know $\frac{1^*}{3^*} = .3333\ldots$, and $\frac{1^*}{1^*} = 1^*$, so the claim is $(.3333\ldots) + (.3333\ldots) + (.3333\ldots) = 1^*$ which is easy to verify. More generally,

Theorem 7.21.2 *In base n, let $\frac{a}{b}$ and $\frac{c}{d}$ be fractions. Then $\frac{a^*}{b^*} + \frac{c^*}{d^*}$ is the infinite decimal associated with the rational number $\frac{a}{b} + \frac{c}{d}$.*

Proof By the usual techniques for adding fractions,

$$\frac{a}{b} + \frac{c}{d} = \frac{a \cdot d + b \cdot c}{b \cdot d}$$

But also, as infinite decimals, by Exercise 7.20.3,

$$\begin{aligned}
\frac{a^*}{b^*} + \frac{c^*}{d^*} &= \frac{(a^*) \cdot (d^*) + (b^*) \cdot (c^*)}{(b^*) \cdot (d^*)} \\
&= \frac{(a \cdot d)^* + (b \cdot c)^*}{(b \cdot d)^*} \\
&= \frac{(a \cdot d + b \cdot c)^*}{(b \cdot d)^*}
\end{aligned}$$

and this is the infinite decimal we have associated with $\frac{a \cdot d - b \cdot c}{b \cdot d}$, that is, with $\frac{a}{b} + \frac{c}{d}$. □

Similar results must be established for multiplication, order, subtraction, division. These are left as exercises, and involve the results of Section 20. For exercises 7.21.2 through 7.21.5 the setting is base n, and $\frac{a}{b}$ and $\frac{c}{d}$ are fractions.

Exercises

Exercise 7.21.1 Use the definition of division for infinite decimals and show, in base 10, $\frac{2^*}{6^*} = .3333\ldots$.

Exercise 7.21.2 Show $\frac{a^*}{b^*} \cdot \frac{c^*}{d^*}$ is the infinite decimal associated with the rational number $\frac{a}{b} \cdot \frac{c}{d}$.

Exercise 7.21.3 Show $\frac{a^*}{b^*} > \frac{c^*}{d^*}$ if and only if $\frac{a}{b} > \frac{c}{d}$.

Exercise 7.21.4 Suppose $\frac{a}{b} \geq \frac{c}{d}$. Show $\frac{a^*}{b^*} - \frac{c^*}{d^*}$ is the infinite decimal associated with the rational number $\frac{a}{b} - \frac{c}{d}$.

Exercise 7.21.5 Suppose $\frac{c}{d} \neq 0$. Show $\frac{a^*}{b^*} \div \frac{c^*}{d^*}$ is the infinite decimal associated with the rational number $\frac{a}{b} \div \frac{c}{d}$.

Finally, we have an important result that essentially says an infinite decimal is determined completely if we know what rational numbers it is bigger than.

Exercise 7.21.6 In base n, let A be an infinite decimal, and let $\mathbf{C}$ be the collection of all infinite decimals which are: 1) associated with rational numbers and 2) are smaller than A. Prove that A is the least upper bound of $\mathbf{C}$. Hint: among those infinite decimals corresponding to rationals are the finite decimals.

7.22   Computing divisions

We know that, provided $B \neq 0^*$, $\frac{A}{B}$ is meaningful. But our proof of this in Section 19 is of no direct help if we want to compute $\frac{A}{B}$ for particular choices of A and B. In this section we develop techniques for actually producing better and better approximations to $\frac{A}{B}$.

Theorem 7.22.1 *In base n, let A and B be infinite decimals with $B \neq 0^*$. We define a sequence of terminating decimals as follows.*
Let d_0 be the largest whole number such that $A \geq B \cdot (d_0)^*$.
Let d_1 be the largest one-place decimal such that $A \geq B \cdot (d_1)^*$.
Let d_2 be the largest two-place decimal such that $A \geq B \cdot (d_2)^*$.
Etc.
Then the sequence $(d_0)^*$, $(d_1)^*$, $(d_2)^*,\ldots$ has a limit, and that limit is $\frac{A}{B}$.

Proof By the first two exercises below, we know the sequence $(d_0)^*$, $(d_1)^*$, $(d_2)^*,\ldots$ has a limit, call it L. By the second of these exercises, $\frac{A}{B} \geq (d_i)^*$ for each i, so $\frac{A}{B} \geq L$. We claim $\frac{A}{B} = L$.

Suppose not, that is, suppose $\frac{A}{B} > L$. Then there is a terminating decimal T between them, $\frac{A}{B} > T > L$. T is associated with some finite decimal, say d. That is, $T = (d)^*$, so $\frac{A}{B} > (d)^* > L$. Let us say d is a k-place finite decimal. Since $\frac{A}{B} > (d)^*$, $\frac{A}{B} \cdot B > (d)^* \cdot B$ or $A > B \cdot (d)^*$. But, d_k is the *largest* k-place finite decimal such that $A \geq B \cdot (d_k)^*$, so $d_k \geq d$, and so $(d_k)^* \geq (d)^* > L$. But also, L, being the limit of a non-decreasing sequence, is also its least upper bound, so $L \geq (d_k)^*$. This is a contradiction, so it must be that $\frac{A}{B} = L$. □

This theorem provides us with a concrete sequence $(d_0)^*$, $(d_1)^*$, $(d_2)^*,\ldots$ of better and better approximations to $\frac{A}{B}$, and Exercise 7.22.3 will give us a technique to calculate them. According to Exercise 7.22.3, having calculated d_k, in order to calculate d_{k+1}, a better approximation to $\frac{A}{B}$, we don't need to begin all over again. We only need to determine the last decimal place of d_{k+1}; we already know all the others since we know what d_k is. And, in base n, there are only n possibilities for that last decimal place, so we can determine it by trial and error.

For example, consider the base 10 problem:

$$\frac{.63237528\ldots}{.20301764\ldots} = \frac{A}{B}.$$

Suppose, using the technique outlined above, we have already produced a one-place approximation to $\frac{A}{B}$. That is, we have found the largest one-place decimal d_1 such that $A \geq B \cdot (d_1)^*$. In fact, $d_1 = 3.1$. Now we want a better, two-place approximation to $\frac{A}{B}$. We want the largest two-place decimal d_2 such that $A \geq B \cdot (d_2)^*$. By Exercise 7.22.3 we know $d_2 = 3.2_-$, and there are only 10 possible digits to fill that blank.

By guesswork we decide to try 2 first. But $B \cdot (3.12)^* = (.20301764\ldots) \cdot (3.120000\ldots) = .6334\ldots > .6323\ldots = A$ so 2 is too big.

So we try 1 next. This time $B \cdot (3.11)^* = (.20301764\ldots) \cdot (3.110000\ldots) = .6316\ldots < .6323\ldots = A$, so 1 is the correct last digit. d_2 is 3.11, and $(3.11)^*$ is a better approximation to $\frac{A}{B}$ than $(3.1)^*$ was.

This technique of approximating to a division is easiest in base 2 as there are only two digits to try.

Exercises

Exercise 7.22.1 Show $(d_0)^*$, $(d_1)^*$, $(d_2)^*$, ... is non-decreasing.

Exercise 7.22.2 Show $\frac{A}{B}$ is an upper bound for the sequence $(d_0)^*$, $(d_1)^*$, $(d_2)^*$, ...

Exercise 7.22.3 In the sequence $(d_0)^*$, $(d_1)^*$, $(d_2)^*$, ..., show the following. All terms have the same whole number parts. All terms from $(d_1)^*$ on have the same first decimal places. All terms from $(d_2)^*$ on have the same second decimal places. And so on.

Exercise 7.22.4 In base 2, work out a four-place approximation to

$$\frac{11.0110111001\ldots}{1.1101011000111\ldots}.$$

7.23   Repeating decimals and rational numbers

We know there are infinite decimals that behave like rational numbers. In this section we see how easy they are to produce and to recognize.

We begin with a base 10 example. Consider the fraction $\frac{3}{11}$. In Section 4 of Chapter 6 we gave a method of approximating to this arbitrarily closely by finite decimals, using long division. For instance, if we want a 5-place approximation we compute:

$$
\begin{array}{r}
.27272 \\
11)\overline{3.00000} \\
\underline{22} \\
80 \\
\underline{77} \\
30 \\
\underline{22} \\
80 \\
\underline{77} \\
30
\end{array}
$$

Now, .27272 is not $\frac{3}{11}$, as a matter of fact, $\frac{3}{11} > .27272$. But the discussion in Chapter 6 shows .27272 is the best 5-place decimal approximation to $\frac{3}{11}$ from underneath. That is, .27272 is the largest 5-place decimal d satisfying $3 \geq 11 \cdot d$.

So much for finite decimals. Now we turn to the problem of approximating to the infinite decimal $\frac{3^*}{11^*}$, which is the counterpart of the rational number $\frac{3}{11}$. Suppose we try the technique of the last section to produce a sequence of better and better approximations to the infinite decimal $\frac{3^*}{11^*}$. Using the notation from that section, d_k is the largest k-place decimal such that $(3)^* \geq (11)^* \cdot (d_k)^*$. Then by our discussion above, d_5 must be .27272. And since all terms in the sequence $(d_0)^*$, $(d_1)^*$, $(d_2)^*$, ... from $(d_5)^*$ on have the same first decimal places, we know that the infinite decimal $\frac{3^*}{11^*}$ must begin .27272....

This discussion is clearly a general one. We can compute as many places of $\frac{3^*}{11^*}$ as we want by carrying out the finite decimal division $11)\overline{3.00...0}$ 'far enough'. And similarly for any other infinite decimal corresponding to a rational number.

Going back to the example above, $\frac{3^*}{11^*}$ in base 10. We found that, written as an infinite decimal it began .27272.... There seems to be a regular pattern, namely 27, that repeats. More generally,

Definition 7.23.1 In base n, if an infinite decimal, after some initial string of decimal places, consists of a block of digits repeated over and over, it is called a *repeating decimal*.

For example, the following are repeating decimals:

$$28.613247247247247\ldots$$
$$18.333333\ldots$$
$$7.6500000\ldots$$

It looks like $\frac{3^*}{11^*}$ is a repeating decimal. In fact it is, and something much more general is true.

Theorem 7.23.2 *In base n, any infinite decimal which corresponds to a rational number is a repeating decimal.*

Rather than prove this in detail, we give an example which will illustrate all the ideas of a proof quite well. Let us show that, in base 10, $\frac{(18)^*}{(241)^*}$ must be a repeating decimal. Well, as we have seen, we may generate this infinite decimal by computing more and more places of

$$241)\overline{18.0000\ldots}$$

This computation begins

$$
\begin{array}{r}
.07 \\
241)\overline{18.0000\ldots} \\
\underline{16\ 87} \\
1\ 13
\end{array}
$$

and at this stage we have a remainder of 113. After the next step we have

$$
\begin{array}{r}
.074 \\
241)\overline{18.0000\ldots} \\
\underline{16\ 87} \\
1\ 130 \\
\underline{964} \\
166
\end{array}
$$

At this stage we have a remainder of 166. And so on. Now, at each step the remainder must be smaller than 241. Since there are only 241 whole numbers smaller than 241 (counting 0), after at most 241 steps we must repeat an earlier remainder. And it is easy to see that, having repeated an earlier remainder the entire calculation from that point on must repeat. Thus we get a periodic decimal, and we know it must repeat after at most 241 terms.

In fact, the converse of this theorem is also true.

Theorem 7.23.3 *In base n, a repeating decimal is one which corresponds to a rational number.*

Once again, rather than give a detailed proof, we give an example which will illustrate the ideas of such a proof. We show that in base 10, 7.282828... corresponds to a rational number.

Let us set $A = 7.282828\ldots$ This repeats with a 'cycle' of 2, so if we move the decimal point two places to the right we get an infinite decimal with the same decimal part. We know multiplying by $(10)^*$ moves a decimal point one place to the right, so multiplying by $(100)^* = (10)^* \cdot (10)^*$ moves it two places. Thus $(100)^* \cdot A = 728.282828\ldots$. Now if we subtract $(100)^* \cdot A - A$, the decimal parts cancel:

$$
\begin{array}{r}
728.282828\ldots \\
-\qquad 7.282828\ldots \\
\hline
721.000000\ldots
\end{array}
$$

So

$$
\begin{aligned}
(100)^* \cdot A - A &= (721)^* \\
(100)^* \cdot A - (1)^* \cdot A &= (721)^* \\
[(100)^* - (1)^*] \cdot A &= (721)^* \\
(99)^* \cdot A &= (721)^* \\
A &= \frac{(721)^*}{(99)^*}
\end{aligned}
$$

so A corresponds to a rational number, $\frac{721}{99}$.

We now know precisely what infinite decimals correspond to rational numbers: they are the repeating decimals. So it is easy to produce an example of an infinite decimal that does *not* correspond to a rational number, for example .10100100010000100000100000001..... In this, there are longer and longer strings of 0's, but there is no fixed block that repeats. Also, we showed in Section 1 that $\sqrt{2}$ was not a rational number. It follows that if there is an infinite decimal whose square is $(2)^*$, it can not be a repeating decimal.

From now on we are going to drop the distinction between rational numbers and the infinite decimals that correspond to them. We will simply write $\frac{2}{3}$, say, and mean ambiguously the rational number named by this fraction, and the infinite decimal $\frac{2^*}{3^*}$. This will cause no problems, and will make reading easier.

Exercises

Exercise 7.23.1 In base 2, by the methods above, compute $\frac{(11)^*}{(101)^*}$ to 6 places.

Exercise 7.23.2 In base 2, what repeating decimal is $\frac{(1)^*}{(11)^*}$?

Exercise 7.23.3 Translate the base 2 fraction $\frac{1}{11}$ into base 10, then find a base 10 repeating decimal to correspond to it.

Exercise 7.23.4 Find a fraction to correspond to .100100100100... in base 2.

Exercise 7.23.5 Consider the infinite decimal $(.110110110110\ldots)_2$. Find a base 10 counterpart. That is, find a fraction, written in base 2 notation, associated with $(.110110110110\ldots)_2$, convert that fraction into base 10 notation, then find the base 10 infinite decimal associated with it.

7.24   Translating from one base to another

We have been developing systems of infinite decimals for all bases. But, all of these systems are equivalent. We already know how to translate from one base to another for certain kinds of infinite decimals, the repeating ones, since these are rational numbers which are available in any system of infinite decimals. Now we extend that translation in a natural way to cover all infinite decimals, repeating or not.

We know how to go from one base to another for rational numbers. Now suppose we don't have a rational. What might we do? As an example, we try to produce a reasonable counterpart in base 10 for the base 2 infinite decimal

$$(.101101110111101\ldots)_2$$

Whether or not this is a rational number, it is the limit of a sequence of rational numbers, $(.1)^*$, $(.10)^*$, $(.101)^*$, $(.1011)^*$, etc. And each term in this sequence has its base 10 counterpart. Translating these into base 10 gives us the following.

$$(.1)^*_2 = (\frac{1}{2})_2 = (\frac{1}{2})_{10} = (.500000000000\ldots)_{10}$$

$$(.10)^*_2 = (\frac{10}{100})_2 = (\frac{2}{4})_{10} = (.500000000000\ldots)_{10}$$

$$(.101)^*_2 = (\frac{101}{100})_2 = (\frac{5}{8})_{10} = (.625000000000\ldots)_{10}$$

$$(.1011)^*_2 = (\frac{1011}{10000})_2 = (\frac{11}{16})_{10} = (.6871500000\ldots)_{10}$$

etc.

We are producing a sequence of base 10 infinite decimals. It is easy to see this is a non-decreasing sequence which is bounded from above (by 1, say), so it has a limit, L. We haven't carried the calculations out far enough yet to say much about what the base 10 infinite decimal L looks like. Still, L exists, and it is quite reasonable to call L the base 10 counterpart of $(.1011011101\ldots)_2$.

So we have a very natural way of translating this particular base 2 infinite decimal into base 10. And it should be clear that our discussion has really been quite general, and would work for any infinite decimal from any base to any base.

Definition 7.24.1 Let $A = w_A.a_1a_2a_3a_4\ldots$ be a base n infinite decimal. We define the *base k translate* of A as follows. First, construct the following sequence of base n terminating decimals.

$$A_0 = (w_A)^*$$
$$A_1 = (w_A.a_1)^*$$
$$A_2 = (w_A.a_1a_2)^*$$
etc.

These, being rational numbers, have their counterparts in base k. Let A'_i be the base k counterpart of A_i. Let A' be the limit of the sequence of base k infinite decimals A'_0, A'_1, A'_2, $A'_3,\ldots$. We define A' to be the base k translate of A.

This is not the only way a translation from base to base can be defined. Here is a second approach.

Alternate definition Let A be a base n infinite decimal, and let $\mathbf{C}$ be the collection of all base n infinite decimals which are rationals smaller than A. (A is the least upper bound of $\mathbf{C}$). Let $\mathbf{C}'$ be the collection of base k counterparts for the members of $\mathbf{C}$. (It is easy to see $\mathbf{C}'$ is bounded from above and so has a least upper bound.) Let A' be the least upper bound of $\mathbf{C}'$ in the system of base k infinite decimals. We define A' to be the base k translate of A.

The equivalence of the two definitions is left as a series of exercises.

We now have a systematic way of translating infinite decimals from one base to another. By itself, that does not establish the essential equivalence of any two bases; we have yet to see how the translation procedure relates to addition, multiplication and the like. For example, if we are going from base 2 to base 10, does it matter whether we carry out our additions in base 2, then translate the result into base 10, or whether we first translate the infinite decimals being added into base 10, and carry out the additions there? In fact, we get the same result either way. And a similar thing happens for subtraction, multiplication and division. But in order to prove this conveniently we need a more general notion of limit than we have been using. Introducing it and developing its properties would take us too far afield, so we content ourselves by saying, "this can be shown." Nothing that follows depends on it.

Exercises

Exercise 7.24.1 Find a base 2 counterpart for the base 10 repeating decimal $(.16161616\dots)_{10}$. Do this by converting the decimal into a base 10 fraction, that into a base 2 fraction, and that into a base 2 infinite decimal.

Exercise 7.24.2 The translation procedure from base 2 to base 10 will work, of course, even if the infinite decimal happens to be rational. Re-do the previous exercise using the technique we outlined for infinite decimals.

Exercise 7.24.3 Prove in Definition 7.24.1 that A'_0, A'_1, A'_2, $A'_3, \dots$ has a limit.

Exercise 7.24.4 In base n, let $\mathbf{C}$ be the collection of all rationals smaller than L, and let A_1, A_2, A_3, $A_4, \dots$ be a bounded, non-decreasing sequence having L as limit. Show that for each B in $\mathbf{C}$ there is a term of the sequence, A_k, such that $B > A_k$.

Exercise 7.24.5 In base n, let $\mathbf{C}$ be a collection of rationals having C as least upper bound, and let A_1, A_2, A_3, $A_4, \dots$ be a non-decreasing sequence of rationals having L as limit. Suppose each A_k is in $\mathbf{C}$ and for each B in $\mathbf{C}$ there is a term of the sequence A_k such that $B > A_k$. Prove $C = L$.

Exercise 7.24.6 Prove the two definitions of translation give the same results.

7.25 Square roots

We saw in Section 21 that the system of base n infinite decimals can be thought of as extending the system of rational numbers. We also saw in Section 23 that there were infinite decimals (the non-repeating ones) that did not correspond to any rational number, hence we have a strictly larger system. But even so, do we have enough? This Chapter began by showing an inadequacy of the rational number system, there was no square root for 2. Well, is there one in the larger system of infinite decimals?

In this section we show that every infinite decimal has a square root, and we discuss briefly how to compute it. This is as far as we take things. But by a similar though more complicated argument one can show that k^{th} roots always exist for any $k = 2, 3, 4, \ldots$. Once this is known, it is possible to develop a general theory of exponents, something we were also able to do in the whole number system, though not in the rational one. But this is further than we wish to carry things. We stop after square roots.

Theorem 7.25.1 *In base n, let A be an infinite decimal. There is an infinite decimal B such that $B^2 = B \cdot B = A$.*

Proof Form a collection $\mathbf{C}$ of base n infinite decimals as follows: put X in $\mathbf{C}$ just when $X^2 \leq A$. That is,

$$\mathbf{C} = \{X \mid X^2 \leq A\}$$

Since $0 \cdot 0 = 0 \leq A$ then $\mathbf{C}$ contains 0 and so is not empty.

Next we show $\mathbf{C}$ has an upper bound. More precisely, we show $A + 1$ is an upper bound. Well, suppose $X \in \mathbf{C}$. If we had $X > A + 1$ then we would have

$$
\begin{aligned}
X \cdot X \quad &> \quad (A+1) \cdot (A+1) \\
&> \quad A \cdot (A-1) \quad \text{(by Theorems 7.17.1 and 7.17.3)} \\
&= \quad A \cdot A + A \cdot 1 \\
&= \quad A \cdot A + A \\
&\geq \quad A
\end{aligned}
$$

that is, $X^2 > A$ which is impossible since $X \in \mathbf{C}$. Thus if $X \in \mathbf{C}$ then $X \leq A + 1$ so $\mathbf{C}$ has an upper bound.

Then by Theorem 7.9.7, $\mathbf{C}$ has a *least* upper bound, call it B. We claim $B^2 = A$, and we show this by deriving contradictions from both $B^2 < A$ and $B^2 > A$.

Case 1: Suppose we had $B^2 < A$. Then $A - B^2$ is meaningful, and is not 0. Also $2 \cdot B + 1 \geq 1 > 0$ so $2 \cdot B + 1$ is not 0. Then by Lemma 7.19.2 there is an infinite decimal $K \neq 0$ such that $K \cdot (2 \cdot B + 1) < (A - B^2)$. Now if we have some choice of K that satisfies this inequality, so would any smaller choice, so we can always arrange things so that K is smaller than 1 as well. We suppose this has been done, and thus $0 < K < 1$. Now consider the infinite decimal

$B + K$. Since $K > 0$, $B + K > B$. Since B is the least upper bound for **C**, this says $B + K$ is *not* a member of **C**. On the other hand,

$$
\begin{aligned}
(B+K)^2 &= B^2 + 2 \cdot B \cdot K + K^2 \\
&= B^2 + K \cdot (2 \cdot B + K) \\
&< B^2 + K \cdot (2 \cdot B + 1) \quad \text{(since } K < 1\text{)} \\
&< B^2 + (A - B^2) \\
&= A
\end{aligned}
$$

which says $B + K \in$ **C**. We have reached a contradiction. Thus we do not have $B^2 < A$.

Case 2: Suppose we had $B^2 > A$. Then (following the proof of Theorem 7.19.3) we proceed as follows.

Since $B \cdot B > A$ there is a terminating decimal T between them, $B \cdot B > T > A$. By Lemma 7.13.4 there are terminating decimals b and c with $B > b$, $B > c$, and $b \cdot c > T$. Let d be the larger of b and c. Then $B > d$ and

$$
\begin{aligned}
d^2 &= d \cdot d \\
&\geq b \cdot c \\
&> T \\
&> A
\end{aligned}
$$

Since $d^2 > A$ it follows that d is an upper bound for **C** (show this). But $d < B$ and B is the *least* upper bound for **C** so d could not be an upper bound. This contradiction says we do not have $B^2 > A$.

The proof is complete. $\square$

Next we turn to the problem of actually computing square roots. This is modeled on the work in Section 22.

Theorem 7.25.2 *In base n, let A be an infinite decimal. We define a sequence of terminating decimals as follows.*

Let d_0 be the largest whole number such that $(d_0^*)^2 \leq A$.
Let d_1 be the largest one-place decimal such that $(d_1^*)^2 \leq A$.
Let d_2 be the largest two-place decimal such that $(d_2^*)^2 \leq A$.
Etc.
Then the sequence $d_0^*, d_1^*, d_2^*, \ldots$ has a limit, and that limit is the square root of A. Further, in this sequence all terms have the same whole number parts; all terms from d_1^* on have the same first decimal places. All terms from d_2^* on have the same second decimal places. And so on.

We leave the proof of this theorem to you as a series of exercises. By using this theorem we can calculate square roots as follows. Say we want, in base 10, the square root of 2. We begin by finding the largest whole number d_0 whose square is ≤ 2. Clearly this is 1. Next we find the largest one-place decimal d_1 whose square is ≤ 2. But d_1 has the same whole number part as d_0, hence all we need do is fill in the blank in $d_1 = 1.\_$. We can use trial-and-error, since there are only 10 possibilities. A little work shows $d_1 = 1.4$. Then d_2 must be of the form 1.4_ and we can fill the blank here by yet another trial-and-error search.

By these means we can compute approximations to the square root of 2 to however many decimal places we desire. In practice there are much better methods of computing square roots. This is not the place to go into them. We have shown the existence of at least one method.

Exercises

Exercise 7.25.1 Show that square roots are unique. That is, suppose $X^2 = A$ and $Y^2 = A$ and show $X = Y$.

Exercise 7.25.2 Show d_0^*, d_1^* d_2^*, ... is non-decreasing in Theorem 7.25.2.

Exercise 7.25.3 Let B be the square root of A. Show B is an upper bound for the sequence d_0^*, d_1^*, d_2^*, ... using the notation of Theorem 7.25.2.

Exercise 7.25.4 Using the notation of Theorem 7.25.2, by the two previous exercises the sequence d_0^*, d_1^*, d_2^*, ... has a limit. Show that the limit is the square root of A.

Exercise 7.25.5 Still using the notation of Theorem 7.25.2, for the sequence d_0^*, d_1^*, d_2^*, ..., show all terms have the same whole number parts; all terms from d_1^* on have the same first decimal places; etc.

Exercise 7.25.6 In base 2 compute a 4-place approximation to the square root of $(10)_2 = (2)_{10}$.

7.26 Real numbers

We now have carried the development of the system of base n infinite decimals quite far. We have established many general properties, and we know any two bases are intertranslatable.

From now on we no longer specify what base we are working in, because from now on it doesn't matter. We are interested only in general structural properties common to all bases (like commutativity of addition) not in computational pecularities (such as whether $1 + 1$ is denoted 2 or 10). To emphasize this point, from now on we will rarely use the term 'infinite decimal'. Rather we will talk of *real numbers*. You may think of the system of real numbers as being the system of base n infinite decimals for some fixed choice of n. Pick your favorite value. If we have examples to give, we will give them using the conventional choice, base 10.

CHAPTER 8

Signed Numbers

Her age, upon the date
Of his birth, was minus eight,
If she's seventeen and he is five-and-twenty!

Iolanthe
W. S. Gilbert, 1882

8.1 Introduction

We have seen number systems developed in which one can express how many, and how long. Now we introduce *signed* numbers, which are used to express, by how much did something change, understanding that changes can be increases or decreases.

There are several places in this book where signed numbers could have been introduced. All that was needed was a 'suitable' number system that included 0. Thus signed numbers could have been introduced after Chapter Three on whole numbers, or after Chapter Five on rational numbers, or after Chapter Seven on real numbers. Consequently we have written this to some extent as a 'floating' chapter. You may think of it as coming after whichever of those three chapters you like.

Throughout this chapter we use the ambiguous term *number*. You may read it systematically as meaning *whole* number, or as meaning *rational* number, or as meaning *real* number. The particular choice does not matter, as we said, but make one and keep it in mind. When we narrow things down to *real* numbers, we will say so.

8.2 Signed numbers

There are many reasons why signed numbers are useful. One to keep in mind for motivation in the next few sections is that it is important to have a system in which *changes* in quantity, as well as absolute amounts, can be easily expressed.

Let us imagine we are talking about the ocean, or a mountain of sand, or the national debt, or whatever. The item should be large enough so that our personal alterations will not exhaust it. Now, suppose you are told: increase it by this, decrease it by that, do thus and so. Then you are asked: by how much have you changed it? We want a number system in which such calculations can be easily carried out.

What we need are two distinct copies of our number system, one representing instructions to increase something by that amount, the other representing instructions to decrease. But no change at all is both an increase and a decrease by 0, which means 0 should belong to both groups of instructions. This is awkward, since we want to be clear, in every instance, whether we are talking about an increase or decrease. It is best, then, to create a third category, just for 0. Thus, what we really want is a system in which there are two copies of the *non-zero* numbers, one intuitively representing increases, the other, decreases, and also there is 0, representing no change.

In order to symbolically represent our two copies of the collection of non-zero numbers we introduce two symbols, $\mathcal{I}$ and $\mathcal{D}$ which can be used as prefixes. If X is a non-zero number, we can think of $\mathcal{I}X$ as telling us: increase whatever is being talked about by X, and we may think of $\mathcal{D}X$ as telling us: decrease it by X. Please note: $\mathcal{I}X$ is *not* $\mathcal{I}$ times X. $\mathcal{I}$ is not a number. Intuitively, $\mathcal{I}$ and $\mathcal{D}$ are instructions.

Definition 8.2.1 By a *signed number* we mean $\mathcal{I}X$ or $\mathcal{D}X$ where X is a non-zero number, or we mean 0.

In order to emphasize the distinction, we will often call the numbers of earlier chapters *unsigned* numbers. We use the informal convention of letting capitals, X, Y, Z, etc. represent unsigned numbers. From now on we will systematically use small letters, x, y, z etc. to represent signed numbers. Thus if we say: let x be a signed number, we mean: either x is 0, or there is a non-zero unsigned number X and x is one of $\mathcal{I}X$ or $\mathcal{D}X$.

Definition 8.2.2 A signed number is called *positive* if it is of the form $\mathcal{I}X$ and *negative* if it is of the form $\mathcal{D}X$.

Thus positive signed numbers intuitively represent increases, and negative signed numbers, decreases.

Important Observation Every signed number is either positive, zero, or negative. No signed number fits into more than one of these categories.

Next we define an operation of changing the sign.

Definition 8.2.3 Let x be a signed number. We define $-x$ as follows.

$$-x = \begin{cases} \mathcal{D}X & \text{if } x = \mathcal{I}X \\ \mathcal{I}X & \text{if } x = \mathcal{D}X \\ 0 & \text{if } x = 0 \end{cases}$$

Intuitively, whatever x tells you to do, $-x$ tells you to do the opposite.

Exercises

Exercise 8.2.1 Show $-(-x) = x$.

Exercise 8.2.2 Show that $x = -x$ if and only if $x = 0$.

8.3   Addition

A definition of addition for signed numbers is easy to motivate, even if lengthy to state in full. Remember the underlying idea that positive numbers represent instructions to increase, negative numbers, to decrease.

Suppose you are told: increase by 2, then increase again by 5. Clearly this gives an increase of $2 + 5$, or 7. Then $\mathcal{I}2 + \mathcal{I}5$ ought to be $\mathcal{I}(2 + 5)$, or $\mathcal{I}7$.

Suppose you are told: increase by 2, then decrease by 7. Clearly the net effect is a decrease by 5. That is, $\mathcal{I}2 + \mathcal{D}7$ ought to be $\mathcal{D}(7 - 2)$, or $\mathcal{D}5$.

Considerations like these lead to the definition below. It has many parts since x can be positive, 0, or negative, and similarly for y. Thus it is simplest to give it partly in chart form.

Definition 8.3.1 Let x and y be signed numbers. We define $x + y$ as follows.
First, if one of x or y is 0:

$$x + 0 = x$$

$$0 + y = y.$$

Next, if neither x nor y is 0:

if x is	and y is	then $x + y$ is	
$\mathcal{I}X$	$\mathcal{I}Y$	$\mathcal{I}(X + Y)$	
$\mathcal{D}X$	$\mathcal{D}Y$	$\mathcal{D}(X + Y)$	
$\mathcal{I}X$	$\mathcal{D}Y$	$\begin{cases} \mathcal{I}(X - Y) & \text{if } X > Y \\ 0 & \text{if } X = Y \\ \mathcal{D}(Y - X) & \text{if } Y > X \end{cases}$	
$\mathcal{D}X$	$\mathcal{I}Y$	$\begin{cases} \mathcal{D}(X - Y) & \text{if } X > Y \\ 0 & \text{if } X = Y \\ \mathcal{I}(Y - X) & \text{if } Y > X \end{cases}$	

Thus the general pattern is the familiar: if the two signs are the same, add the unsigned numbers and give the result the common sign; if the signs are different, subtract the unsigned numbers, whichever way makes sense, and give the result the sign of the larger.

Since the definition of addition is in several cases, we may expect proofs of theorems on addition to involve several cases too. So we begin with a result that may be used to reduce the number of cases that need be considered.

Theorem 8.3.2 *For any signed numbers x and y, $-(x + y) = (-x) + (-y)$.*

Proof First, if either x or y is 0, the result is immediate.

Now, if neither x nor y is 0, then there are two possibilities for x, positive and negative, and similarly for y. Thus there are $2 \cdot 2$ or 4 cases to consider.

Case 1: x and y both positive, say $x = \mathcal{I}X$ and $y = \mathcal{I}Y$. Then

$$
\begin{aligned}
-(x+y) &= -[\mathcal{I}X + \mathcal{I}Y] \\
&= -[\mathcal{I}(X+Y)] \\
&= \mathcal{D}(X+Y) \\
&= \mathcal{D}X + \mathcal{D}Y \\
&= [-(\mathcal{I}X)] + [-(\mathcal{I}Y)] \\
&= (-x) + (-y).
\end{aligned}
$$

Case 2: x and y both negative, is treated similarly.

Case 3: x positive, y negative, say $x = \mathcal{I}X$ and $y = \mathcal{D}Y$. Then, using Exercise 8.3.2,

$$
\begin{aligned}
-(x+y) &= -[\mathcal{I}X + \mathcal{D}Y] \\
&= \mathcal{D}X + \mathcal{I}Y \\
&= [-(\mathcal{I}X)] + [-(\mathcal{D}Y)] \\
&= (-x) + (-y).
\end{aligned}
$$

Case 4: x negative, y positive, is treated similarly. □

Now we see how this theorem may be used.

Theorem 8.3.3 (Commutativity of addition) $x + y = y + x$.

Since x can be positive, zero or negative, and similarly for y, there would seem to be $3 \cdot 3$ or 9 cases to consider. But, if either x or y is 0 the result is trivial, so we are effectively reduced to $2 \cdot 2$ or 4 cases. Further, we can cut that number in half since, if the result is true for x and y, it is also true for $-x$ and $-y$. This happens since, using Theorem 8.3.2, if we know that $x + y = y + x$, then

$$
\begin{aligned}
(-x) + (-y) &= -(x+y) \\
&= -(y+x) \\
&= (-y) + (-x).
\end{aligned}
$$

Thus if we have the theorem for positive x and y, we automatically have it for negative x and y. Similarly if we have it for positive x and negative y, we also have it the other way around. So finally, we are reduced to actually showing the theorem in the following two cases:

1. x and y positive,

2. x positive, y negative.

We leave this to you.

Theorem 8.3.4 (Associativity of addition) $x + (y + z) = (x + y) + z$.

We spend the rest of this section on a proof of Theorem 8.3.4. Most of our discussion will be devoted to reducing the $3 \cdot 3 \cdot 3$ or 27 cases to a manageable number.

First, if any of x, y or z is 0, the result is simple. This reduces things to $2 \cdot 2 \cdot 2$ or 8 cases.

Next, by using Theorem 8.3.2 as we did above, these 8 cases can be cut in half.

By Exercise 8.3.4 it suffices to show the truth of the theorem in four cases, specifically we consider: the case where x, y and z are all positive; the three cases where two of x, y and z are positive, one is negative.

Exercise 8.3.5 leaves us with three cases to consider. But, in fact, two of these follow from the third. We show this for one and leave the other as an exercise.

Temporary assumption $x + (y + z) = (x + y) + z$ whenever x and y are positive, z is negative.

Consequence of temporary assumption $x+(y+z) = (x+y)+z$ whenever x is negative, y and z are positive.

Proof of consequence Suppose $x = \mathcal{D}X$, $y = \mathcal{I}Y$ and $z = \mathcal{I}Z$. Then

$$
\begin{aligned}
x + (y + z) &= \mathcal{D}X + (\mathcal{I}Y + \mathcal{I}Z) \\
&= (\mathcal{I}Y + \mathcal{I}Z) + \mathcal{D}X \quad \text{by Theorem 8.3.3} \\
&= (\mathcal{I}Z + \mathcal{I}Y) + \mathcal{D}X \quad \text{by Theorem 8.3.3} \\
&= \mathcal{I}Z + (\mathcal{I}Y + \mathcal{D}X) \quad \text{by Temporary Assumption} \\
&= (\mathcal{I}Y + \mathcal{D}X) + \mathcal{I}Z \quad \text{by Theorem 8.3.3} \\
&= (\mathcal{D}X + \mathcal{I}Y) + \mathcal{I}Z \quad \text{by Theorem 8.3.3} \\
&= (x + y) + z
\end{aligned}
$$

Finally, using Exercise 8.3.6 we are left with one case, that embodied in our temporary assumption. We now establish that.

Lemma 8.3.5 $x + (y + z) = (x + y) + z$ *whenever x and y are positive, z is negative.*

Proof Say $x = \mathcal{I}X$, $y = \mathcal{I}Y$ and $z = \mathcal{D}Z$. There are three cases to consider. These are:

1. $X + Y > Z$

2. $X + Y = Z$

3. $X + Y < Z$.

We show case 1, the most complicated and leave the other two as exercises.
case 1: $X + Y > Z$. Then

$$
\begin{aligned}
(x+y)+z &= (\mathcal{I}X + \mathcal{I}Y) + \mathcal{D}Z \\
&= \mathcal{I}(X+Y) + \mathcal{D}Z \\
&= \mathcal{I}[(X+Y) - Z]
\end{aligned}
$$

So we must show $x + (y+z)$ is also this quantity. And there are three subcases:

Subcase 1a: $Y > Z$. Then

$$
\begin{aligned}
x + (y+z) &= \mathcal{I}X + (\mathcal{I}Y + \mathcal{D}Z) \\
&= \mathcal{I}X + \mathcal{I}(Y-Z) \\
&= \mathcal{I}[X + (Y-Z)]
\end{aligned}
$$

and this is $\mathcal{I}[(X+Y) - Z]$ by properties of subtraction for unsigned numbers.

Subcase 1b: $Y = Z$. Then

$$
\begin{aligned}
x + (y+z) &= \mathcal{I}X + (\mathcal{I}Y + \mathcal{D}Z) \\
&= \mathcal{I}X \\
&= \mathcal{I}[X + (Y-Z)]
\end{aligned}
$$

and by subtraction properties again, this is $\mathcal{I}[(X+Y) - Z]$.

Subcase 1c: $Z > Y$. Then

$$
\begin{aligned}
x + (y+z) &= \mathcal{I}X + (\mathcal{I}Y + \mathcal{D}Z) \\
&= \mathcal{I}X + \mathcal{D}(Z-Y)
\end{aligned}
$$

Now, in case 1, $X + Y > Z$, so in this subcase, $X > Z - Y$, hence this is $\mathcal{I}[X - (Z-Y)]$ and by subtraction properties, this is $\mathcal{I}[(X+Y) - Z]$. This completes case 1.

Exercise 8.3.7 completes the proof of Theorem 8.3.4. $\square$

Thus we have commutativity and associativity of addition for signed numbers, so from now on we will generally drop parentheses in sums, since their exact placement doesn't matter.

Exercises

Exercise 8.3.1 Prove that $x + (-x) = 0$.

Exercise 8.3.2 Show

1. $-[\mathcal{I}X + \mathcal{D}Y] = \mathcal{D}X + \mathcal{I}Y$

2. $-[\mathcal{D}X + \mathcal{I}Y] = \mathcal{I}X + \mathcal{D}Y$.

Do this by considering the three cases: $X > Y$, $X = Y$ and $Y > X$.

Exercise 8.3.3 Prove theorem 8.3.3 by establishing the two cases given above.

Exercise 8.3.4 Show that if $x + (y + z) = (x + y) + z$, then $(-x) + [(-y) + (-z)] = [(-x) + (-y)] + (-z)$.

Exercise 8.3.5 Show that $x + (y + z) = (x + y) + z$ if all of x, y and z are positive.

Exercise 8.3.6 Show the following is also a consequence of the temporary assumption: $x + (y + z) = (x + y) + z$ whenever x and z are positive, y is negative.

Exercise 8.3.7 Complete the proof of Lemma 8.3.5 by considering the other two cases. Note that in case 2, $X + Y = Z$, there are no subcases since $Y < Z$, and in case 3, $X + Y < Z$, there are no subcases since $Y < Z$ (why?).

8.4 Multiplication

An appropriate definition of multiplication for signed numbers is not easy to discover. After all, the underlying idea behind signed numbers is to simplify addition and subtraction. Multiplication does not play any role in this, so our intuitive feelings about how signed numbers 'ought to behave' are not going to get us very far now. Other considerations must be used. One reasonable guide is that the basic laws of multiplication for unsigned numbers should continue to work for signed numbers. In fact, using this idea, we will be able to deduce what the definition of multiplication ought to be.

First, we want the system of signed numbers to be an extension of the system of unsigned numbers. That is, we want certain signed numbers to 'act like' the unsigned numbers did. The most natural candidates are the positive numbers, since the quantity X is also the result of an increase by X, starting from 0. Thus it is plausible that $\mathcal{I}X$ and X should behave alike in their respective number systems. But then, we must have

$$(\mathcal{I}X) \cdot (\mathcal{I}Y) = \mathcal{I}(X \cdot Y)$$

So we have the first part of a proposal for a definition of multiplication.

Considerations like this also suggest $0 \cdot x$ and $x \cdot 0$ should be 0 if x is not negative; they do not directly suggest a value if x is negative. So, we turn to other methods. Suppose we assume multiplication has been defined in such a way that the distributive law holds. Then we would have

$$x \cdot 0 = x \cdot (0 + 0) = x \cdot 0 + x \cdot 0$$

Now, adding $-(x \cdot 0)$ to both sides (and using Exercise 8.3.1) we get $0 = x \cdot 0$. If we assume the commutative law holds we also would have $0 \cdot x = 0$ so we have some more parts of a proposal for a definition of multiplication.

Continuing along these lines, making the same assumptions, and using what we just came up with, we would have

$$
\begin{aligned}
0 &= (\mathcal{I}X)\cdot 0 \\
&= (\mathcal{I}X)\cdot[\mathcal{I}Y + \mathcal{D}Y] \\
&= (\mathcal{I}X)\cdot(\mathcal{I}Y) + (\mathcal{I}X)\cdot(\mathcal{D}Y) \\
&= \mathcal{I}(X\cdot Y) + (\mathcal{I}X)\cdot(\mathcal{D}Y)
\end{aligned}
$$

Now, adding $\mathcal{D}(X\cdot Y)$ to both sides we get $\mathcal{D}(X\cdot Y) = (\mathcal{I}X)\cdot(\mathcal{D}Y)$. Thus two more parts of our definition are

$$(\mathcal{I}X)\cdot(\mathcal{D}Y) = \mathcal{D}(X\cdot Y)$$

$$(\mathcal{D}X)\cdot(\mathcal{I}Y) = \mathcal{D}(X\cdot Y)$$

Finally Exercise 8.4.1 gives us the last part and we are led to the following.

Definition 8.4.1 Let x and y be signed numbers. We define $x\cdot y$ as follows. First, if one of x or y is 0:

$$x\cdot 0 = 0$$

$$0\cdot y = 0$$

Next, if neither x nor y is 0:

if x is	and y is	then $x\cdot y$ is
$\mathcal{I}X$	$\mathcal{I}Y$	$\mathcal{I}(X\cdot Y)$
$\mathcal{I}X$	$\mathcal{D}Y$	$\mathcal{D}(X\cdot Y)$
$\mathcal{D}X$	$\mathcal{I}Y$	$\mathcal{D}(X\cdot Y)$
$\mathcal{D}X$	$\mathcal{D}Y$	$\mathcal{I}(X\cdot Y)$

Thus we have the familiar, two negatives make a positive, etc. We have seen that if the basic laws of multiplication are to hold, the definition must be as given above. Now we must check whether, given this definition, the laws in fact do hold.

Theorem 8.4.2 (Commutativity of multiplication)
$x\cdot y = y\cdot x$

Proof First, if either x or y is 0, the result is immediate.

Next, there are four cases in which neither x nor y is 0. We do one and leave the other three as exercises.

Suppose x is positive, y is negative, say $x = \mathcal{I}X$ and $y = \mathcal{D}Y$. Then $x\cdot y = (\mathcal{I}X)\cdot(\mathcal{D}Y) = \mathcal{D}(X\cdot Y)$ and by commutativity of multiplication for *unsigned* numbers, this in turn $= \mathcal{D}(Y\cdot X) = (\mathcal{D}Y)\cdot(\mathcal{I}X) = y\cdot x$.

We leave the rest as an exercise. □

Theorem 8.4.3 (Associativity of multiplication) $x\cdot(y\cdot z) = (x\cdot y)\cdot z$.

From now on we will freely omit parentheses in products of three or more signed numbers since, by the two theorems above, their exact placement doesn't matter.

Theorem 8.4.4 (Distributive law) $x \cdot (y + z) = x \cdot y + x \cdot z$.

We conclude this section by showing that some of the clauses of the definition of multiplication are really special cases of a more general phenomenon. The proofs are rather similar to the informal arguments that led to the definition of multiplication in the first place.

Theorem 8.4.5 $(-x) \cdot y = -(x \cdot y)$.

Proof $0 = [x + (-x)] \cdot y = x \cdot y + (-x) \cdot y$. Now, adding $-(x \cdot y)$ to both sides, we get $-(x \cdot y) = (-x) \cdot y$. $\square$

Exercises

Exercise 8.4.1 Show that if the distributive law holds, and the discussion preceding the definition of multiplication is accepted, then $(\mathcal{D}X) \cdot (\mathcal{D}Y) = \mathcal{I}(X \cdot Y)$.

Exercise 8.4.2 Show $(\mathcal{I}1) \cdot x = x$.

Exercise 8.4.3 Complete the proof of Theorem 8.4.2 by doing the other three cases.

Exercise 8.4.4 Prove Theorem 8.4.3.

Exercise 8.4.5 Prove Theorem 8.4.4

Exercise 8.4.6 Prove $x \cdot (-y) = -(x \cdot y)$.

Exercise 8.4.7 Prove $(-x) \cdot (-y) = x \cdot y$.

8.5 Subtraction and division

Both subtraction and division (when appropriate) are very easy to treat, since they may be reduced to addition and multiplication respectively. Having done this, there is little else to say. We should mention, though, that the ease of subtraction for signed numbers is a little misleading. Actually, subtraction has been built into the definition of addition in Section 3. Many results about subtraction of unsigned numbers were needed to establish the associative law for addition of signed numbers. In short, subtraction has been simplified by shifting the area of complexity to addition.

We first discuss subtraction, then division. As usual, we want $a - b$ to be that quantity which will get us from b to a. That is, we want $a - b$ to be the unique solution of $b + x = a$. But it is easy to produce a solution to this equation, just add $-b$ to both sides, use Exercise 8.3.1 (and the associativity and commutativity of addition) to get:

$$
\begin{aligned}
b + x &= a \\
-b + b + x &= a + (-b) \\
0 + x &= a + (-b) \\
x &= a + (-b).
\end{aligned}
$$

Further, this is the *only* solution. See Exercise 8.5.1.

Definition 8.5.1 For any signed numbers a and b, $a - b$ is $a + (-b)$.

Remark Notice that $a - b$ is defined whether a is bigger than b or not. In fact, we have not yet even defined a notion of 'bigger than' for signed numbers.

Next we turn to division. Since exact division was, in general, not possible in the *whole* number system, for the rest of this section, *number* means either *rational* number or *real* number. Then, for each non-zero X, X^{-1} or $\frac{1}{X}$ is meaningful.

Now, one useful way of thinking of $a - b$, or $a + (-b)$ is: it takes us from b to a by addition, and it does so in two stages, first the $-b$ gets us from b to 0, then the a gets us from 0 to a. This suggests an approach to division. $a \div b$ is to be that number which will take us from b to a by multiplication. Now $\mathcal{I}1$ plays the role for multiplication that 0 plays for addition $[x + 0 = x$ and $x \cdot (\mathcal{I}1) = x]$. So perhaps we can get from b to a by multiplication by first getting from b to $\mathcal{I}1$, then it is easy to get from there to a, just multiply by a.

Definition 8.5.2 Suppose $x \neq 0$. We define x^{-1} as follows: if $x = \mathcal{I}X$ then $x^{-1} = \mathcal{I}(\frac{1}{X})$; if $x = \mathcal{D}X$ then $x^{-1} = \mathcal{D}(\frac{1}{X})$.

Remark x^{-1} can not be defined so that Exercise 8.5.4 holds if x is 0 since $0 \cdot y = 0$ for any y. Exercise 8.5.5 serves as motivation for the following definition.

Definition 8.5.3 If $y \neq 0$, $x \div y = x \cdot y^{-1}$. We often write $\frac{x}{y}$ for $x \div y$.

Now that we know what division is for signed numbers, there remains the problem of how best to compute it. Exercise 8.5.7 treats this. And Exercise 8.5.8 generalizes it in certain ways. Finally, analogs of the properties of division given in earlier chapters may be proved. A few are listed as further exercises.

Exercises

Exercise 8.5.1 Show that if $b + x = a$ and also $b + x' = a$, then $x = x'$.

Exercise 8.5.2 Prove the following:

1. $a - b = (a - c) + (c - b)$

2. $(a + b) - c = a + (b - c)$

3. $a - (b + c) = (a - b) - c$

4. $a - (b + c) = (a - c) - b$

5. $a - b = (a + c) - (b + c)$

Exercise 8.5.3 Show that, if $x \neq 0$, $(-x)^{-1} = -(x^{-1})$.

Exercise 8.5.4 Show that if $x \neq 0$, $x \cdot x^{-1} = \mathcal{I}1$.

Exercise 8.5.5 Show that, if $y \neq 0$, $(x \cdot y^{-1}) \cdot y = x$.

Exercise 8.5.6 Suppose $b \neq 0$. Show the equation $b \cdot x = a$ has just one solution, $x = \frac{a}{b}$.

Exercise 8.5.7 Show, if $Y \neq 0$,

1. $\frac{\mathcal{I}X}{\mathcal{I}Y} = \frac{\mathcal{D}X}{\mathcal{D}Y} = \mathcal{I}(\frac{X}{Y})$

2. $\frac{\mathcal{I}X}{\mathcal{D}Y} = \frac{\mathcal{D}X}{\mathcal{I}Y} = \mathcal{D}(\frac{X}{Y})$

Exercise 8.5.8 Suppose $y \neq 0$ and show:

1. $\frac{-x}{-y} = \frac{x}{y}$

2. $\frac{-x}{y} = \frac{x}{-y} = -\frac{x}{y}$

Exercise 8.5.9 If b and d are not 0, show $\frac{a}{b} = \frac{c}{d}$ if and only if $a \cdot d = b \cdot c$.

Exercise 8.5.10 Show that if b and c are not 0:

1. $(b \cdot c)^{-1} = b^{-1} \cdot c^{-1}$

2. $\frac{a}{b} = \frac{a \cdot c}{b \cdot c}$

Exercise 8.5.11 If $x \neq 0$, show $(x^{-1})^{-1} = x$.

Exercise 8.5.12 If $\frac{a}{b} \neq 0$, show $(\frac{a}{b})^{-1} = \frac{b}{a}$.

8.6   A simplified notation

Whenever we introduced a number system it was important to know that earlier number systems were not lost. Well now we have created the system of signed numbers and it is just as important to know the unsigned numbers are still available; there are signed numbers that behave like them. This was foreshadowed in Section 4 when we observed: the quantity X is also the result of an increase by X, starting from 0. In short, the positive numbers together with 0 can be expected to behave like the unsigned numbers did.

So far we have discussed addition, subtraction, multiplication and division, and for each of these operations, it really is the case that the positive numbers and the unsigned numbers behave alike. More precisely, for any unsigned numbers X and Y:

$$\begin{aligned}
\mathcal{I}X + \mathcal{I}Y &= \mathcal{I}(X+Y) \\
(\mathcal{I}X) \cdot (\mathcal{I}Y) &= \mathcal{I}(X \cdot Y) \\
\frac{\mathcal{I}X}{\mathcal{I}Y} &= \mathcal{I}(\frac{X}{Y}) \qquad \text{if } Y \neq 0 \text{ and } \frac{X}{Y} \text{ is defined} \\
\mathcal{I}X - \mathcal{I}Y &= \mathcal{I}(X-Y) \quad \text{if } X > Y
\end{aligned}$$

This similarity of behavior will continue, so we may as well agree to use the same notation for positive numbers that we do for unsigned numbers. As long as we know they are different things that act alike, there is no need to go on emphasizing the distinction symbolically. So from now on we will write, say, $3.1415926535\ldots$ and mean by it, ambiguously, either the unsigned real, $3.1415926535\ldots$, or the signed real $\mathcal{I}3.1415926535\ldots$ Similarly we will write $\frac{2}{3}$ for either the unsigned rational $\frac{2}{3}$ or the signed rational $\mathcal{I}\frac{2}{3}$. This use of $\frac{2}{3}$ is thus doubly ambiguous, since it also stands for a certain real number (and for a particular fraction as well). But again, all these act alike in their respective number systems, so the distinction is harmless. Similar remarks apply to whole numbers too.

A further simplification in notation: $\mathcal{D}X = -(\mathcal{I}X)$ and we are agreeing to write just X for $\mathcal{I}X$, so we may as well write $-X$ for $\mathcal{D}X$. Thus from now on we stop using $\mathcal{D}$ as well as $\mathcal{I}$, writing $-3.1415\ldots$ for $\mathcal{D}3.1415\ldots$.

Important remark $-x$ is not automatically negative. It is negative only if x is positive.

Exercises

Exercise 8.6.1 Prove $\mathcal{I}X - \mathcal{I}Y = \mathcal{I}(X - Y)$ if $X > Y$.

Exercise 8.6.2 Show

1. $x = (1) \cdot x$

2. $-x = (-1) \cdot x$.

8.7   Fields and integral domains

We have now established a number of results about the signed numbers. In fact, most of them may be summarized by saying the systems of signed reals and signed rationals are *fields*, and the system of signed whole numbers is an *integral domain*. Fields and integral domains are standard mathematical structures and much is known about them. In this section we give the definitions and derive some of their elementary properties.

A field or an integral domain is a non-empty collection, $\mathcal{F}$ say, on which is defined two operations meeting certain conditions. Now in the examples we are primarily interested in, $+$ and $\cdot$ are used to denote the two operations so we will use those symbols for an arbitrary field or integral domain as well. But keep in mind, $+$ and $\cdot$ are simply two operations on $\mathcal{F}$. Depending on $\mathcal{F}$, they might be something other than addition and multiplication.

Rather than state all the conditions (axioms) at once, we state a few, discuss them, state a few more, and so on. In our arrangement there are 9 axioms. Most of them are common to both fields and integral domains. You should check as we go along, that the system of signed numbers satisfies each of the axioms.

Axiom 1 $\mathcal{F}$ is closed under $+$ and $\cdot$.

This means that if we apply the operation $+$ to two members of $\mathcal{F}$, the result is a member of $\mathcal{F}$. Similarly for $\cdot$. For example, the positive numbers are closed under multiplication, the negative numbers are not. Of course, the entire set of signed numbers is closed under addition and multiplication as we have defined them.

Axiom 2 (Commutativity) For any members x and y of $\mathcal{F}$,

$$x + y = y + x$$

$$x \cdot y = y \cdot x$$

Axiom 3 (Associativity) For any members x, y and z of $\mathcal{F}$.

$$x + (y + z) = (x + y) + z$$

$$x \cdot (y \cdot z) = (x \cdot y) \cdot z$$

On the basis of the two axioms above, parentheses can be dropped in sums, and likewise in products, and terms in sums or products can be re-ordered. Mixed sums and products are of course, a different matter. They are the subject of the next axiom.

Axiom 4 (Distributivity) For any members x, y and z of $\mathcal{F}$,

$$x \cdot (y + z) = x \cdot y + x \cdot z$$

Axiom 5 (Additive identity) There is a member of $\mathcal{F}$, call it O, satisfying the condition $x + O = x$ for every x in $\mathcal{F}$.

Notice this axiom only says there is an additive identity, it doesn't say there is only one. But in fact that follows rather easily.

Theorem 8.7.1 *In any field or integral domain, there is only one additive identity.*

Proof Suppose O and O' were both additive identities. Then, for every x in $\mathcal{F}$,

    1) $x + O = x$
    2) $x + O' = x$.
Now, let x be O' in 1) and let x be O in 2) , to get:

    $1'$) $O' + O = O'$
    $2'$) $O + O' = O$.
But $O' + O = O + O'$ by Axiom 2, so by $1'$) and $2'$), $O = O'$. $\square$

In the system of signed numbers, *the* additive identity is 0. To keep notation simple, from now on we will use 0 to symbolize the additive identity of our field or integral domain whether or not it is the signed numbers.

Axiom 6 (Multiplicative identity) There is a member of $\mathcal{F}$, call it e, satisfying the condition $x \cdot e = x$ for every x in $\mathcal{F}$.

In the signed numbers, 1 is the multiplicative identity. From now on we follow custom and use the symbol 1 in any field or integral domain instead of e.

All the axioms so far are satisfied by the system of unsigned numbers. The next one is not.

Axiom 7 (Additive inverse) For every x in $\mathcal{F}$ there is some y such that $x + y = 0$.

In Axiom 7, y is called an additive inverse of x. The axiom only says x has an additive inverse; it can be proved it has exactly one.

Lemma 8.7.2 (Cancellation law) *In any field or integral domain, if $a + b = a + c$ then $b = c$.*

Proof Suppose $a + b = a + c$. Let d be any additive inverse for a, so $a + d = 0$. Then $d + (a + b) = d + (a + c)$ so by associativity, $(d + a) + b = (d + a) + c$ and by commutativity, $(a + d) + b = (a + d) + c$; $0 + b = 0 + c$; $b + 0 = c + 0$; and finally $b = c$. $\square$

Theorem 8.7.3 *In any field or integral domain, members have unique additive inverses.*

Proof Suppose x has both y and y' as additive inverses. Then $x + y = 0$ and $x + y' = 0$ so $x + y = x + y'$. Then by the lemma, $y = y'$. □

This theorem allows us to introduce the following notation.

Definition 8.7.4 By $-x$ we mean the unique additive inverse of x.

Remark Since $-x$ is the additive inverse of x, $x + (-x) = 0$. But then, $(-x) + x = 0$ and this says x is the additive inverse of $-x$, or in our notation, $x = -(-x)$.

In Section 3 we showed the result of Exercise 8.7.3 was true of the signed reals before showing commutativity, associativity, or any other properties of addition. We did it that way because it simplified our work in showing those other laws. Now we learn it could have been deduced from them if we had derived them first. The point is, Exercise 8.7.3 is true in any field or integral domain, but actually showing something is a field or integral domain may require detours.

The additive identity, 0, is defined by its behavior under addition; interestingly enough, that determines how it behaves under multiplication as well.

Theorem 8.7.5 *In any field or integral domain, $x \cdot 0 = 0$.*

Proof Using the distributive law, and the fact that 0 is the additive identity,

$$
\begin{aligned}
(x \cdot 0) + 0 &= x \cdot 0 \\
&= x \cdot (0 + 0) \\
&= (x \cdot 0) + (x \cdot 0)
\end{aligned}
$$

Then by the cancellation law, $0 = x \cdot 0$. □

Further, $-x$ is specified by its additive behavior, but it too has a well defined multiplicative role, which we leave to you in Exercise 8.7.4.

Definition 8.7.6 $x - y = x + (-y)$.

We leave the properties of subtraction to you as well.

The next axiom is a *non-triviality* axiom. There are many fields and integral domains besides the signed numbers. The purpose of Axiom 8 is to rule out of consideration *very* degenerate systems.

Axiom 8 $\mathcal{F}$ has more than one member.

Now, finally, fields and integral domains go their separate ways. For rationals and for reals, exact division is always possible, except by 0. That is, every non-zero number has its multiplicative inverse. In the whole number system this is not the case. Still, in the whole number system we do have a cancellation law for multiplication, which carries over to *signed* whole numbers as well.

This allows us to do certain things that we would probably use multiplicative inverses for, if we had them.

The final axiom is stated in two versions, one for fields and one for integral domains.

Field Axiom 9 (Multiplicative inverse) For every x in $\mathcal{F}$ other than 0 there is some y in $\mathcal{F}$ such that $x \cdot y = 1$.

Integral Domain Axiom (Cancellation law) For every z in $\mathcal{F}$ other than 0, if $x \cdot z = y \cdot z$ then $x = y$.

An integral domain is a system satisfying the eight common axioms and the Cancellation law. A field is a system satisfying the eight common axioms and the Multiplicative inverse condition. Note that the signed whole numbers constitute an integral domain, and the signed rationals and signed reals are fields.

Definition 8.7.7 In a field, if $x \neq 0$, by x^{-1} we mean the unique multiplicative inverse of x.

Definition 8.7.8 In a field, if $y \neq 0$, $\frac{x}{y} = x \cdot y^{-1}$.

Exercises

Exercise 8.7.1 Show that in any field or integral domain, $x \cdot [(y + z) + w] = (x \cdot y + x \cdot z) + x \cdot w$.

Exercise 8.7.2 Show that in any field or integral domain there is only one multiplicative identity.

Exercise 8.7.3 Show that in any field or integral domain, $-(x + y) = (-x) + (-y)$.

Exercise 8.7.4 Show that in any field or integral domain,

1. $x \cdot (-y) = -(x \cdot y)$
2. $(-x) \cdot y = -(x \cdot y)$
3. $(-x) \cdot (-y) = x \cdot y$.

Exercise 8.7.5 Show that in any field or integral domain, $-(x - y) = y - x$.

Exercise 8.7.6 Show that in any field or integral domain, $x \cdot (y - z) = x \cdot y - x \cdot z$.

Exercise 8.7.7 Show all parts of Exercise 8.5.2 hold in any field or integral domain.

Exercise 8.7.8 Show that in any field or integral domain, $0 \neq 1$.

Exercise 8.7.9 Show that a cancellation law for multiplication holds in every field. Thus every field is also an integral domain.

Exercise 8.7.10 Show the restriction "other than 0" in the two versions of Axiom 9 is necessary. Hint: show that if it were dropped, Axiom 8 would be violated.

Exercise 8.7.11 Show that in any integral domain, if $x \cdot y = 0$ then $x = 0$ or $y = 0$.

Exercise 8.7.12 Show that in any field, members have unique multiplicative inverses.

Exercise 8.7.13 Show that in any field, if $x \neq 0$, $x = (x^{-1})^{-1}$.

Exercise 8.7.14 Show that in any field, if $x \neq 0$ and $y \neq 0$, then $(x \cdot y)^{-1} = (x^{-1}) \cdot (y^{-1})$.

Exercise 8.7.15 Show that in any field, if $b \neq 0$ and $d \neq 0$:

1. $\frac{a}{b} = \frac{c}{d}$ if and only if $a \cdot d = b \cdot c$;

2. $\frac{a}{b} = \frac{a \cdot d}{b \cdot d}$;

3. $\frac{a}{b} + \frac{c}{d} = \frac{a \cdot d + b \cdot c}{b \cdot d}$;

4. $\frac{a}{b} \cdot \frac{c}{d} = \frac{a \cdot c}{b \cdot d}$.

8.8 An order relation for signed numbers

Producing a satisfactory notion of order is easy. We want $x > y$ to be true if the passage from y to x is an increase. But, by our discussion of subtraction, $x - y$ is what will get us from y to x.

Definition 8.8.1 $x > y$ if $x - y$ is positive.

On the surface, it looks like our definition of $>$ for signed numbers makes no use of the notion of $>$ for unsigned numbers, which we developed at great length in earlier chapters. In fact, this is not the case, though the connection is a hidden one. In order to decide if $x > y$, we must compute $x - y$, that is, $x + (-y)$. Now in order to add signed numbers, if the signs are different, the result gets the sign of the larger unsigned number involved. This means, in order to decide if $x > y$ is true, we need to be able to compare unsigned numbers. In fact, none of our earlier development will be wasted.

Rather than continuing the discussion in its present setting, we go back to fields and integral domains for a more general treatment, having signed numbers as examples.

Exercises

Exercise 8.8.1 Show that unsigned numbers and positive numbers act alike with respect to order. That is, temporarily resuming our old notation, $\mathcal{I}X >\mathcal{I}Y$ in the present sense if and only if $X > Y$ in the sense of earlier chapters.

8.9   Ordered fields and integral domains

Suppose $\mathcal{F}$ is a field or an integral domain using the operations $+$ and $\cdot$. $\mathcal{F}$ is said to be *ordered* if there is a subset $\mathcal{P}$ of $\mathcal{F}$ (intuitively the positive members of $\mathcal{F}$) meeting the following three conditions.

Order axiom 1 $\mathcal{P}$ is closed under $+$.

Order axiom 2 $\mathcal{P}$ is closed under $\cdot$.

Order axiom 3 For every x in $\mathcal{F}$, exactly one of the following holds: x is in $\mathcal{P}$, x is 0, $-x$ is in $\mathcal{P}$.

It should be clear that the signed numbers are ordered (taking $\mathcal{P}$ to be the positive numbers) so anything true of ordered systems in general is true of the signed numbers. In *any* ordered field or integral domain we will use the language x is *positive* to mean x is in $\mathcal{P}$.

Definition 8.9.1 $x > y$ if $x - y \in \mathcal{P}$.

Now the basic properties of $>$ follow rather easily. All the following takes place in an ordered field or integral domain.

Theorem 8.9.2 (Transitivity) *If $x > y$ and $y > z$ then $x > z$.*

Proof Suppose $x > y$ and $y > z$. Then, by definition, $x - y \in \mathcal{P}$ and $y - z \in \mathcal{P}$. But $\mathcal{P}$ is closed under $+$ so $(x - y) + (y - z) \in \mathcal{P}$, that is, $x - z \in \mathcal{P}$. Hence $x > z$. □

This illustrates how things often are proved in ordered fields and integral domains. Translate away all occurrences of $>$, then use the axioms. Indeed we leave the basic properties as exercises. The first several will not need Order Axiom 3. It is assumed in the exercises that we are in an *ordered* field or integral domain.

Theorem 8.9.3 *If $a \neq 0$, a^2 is positive (we are writing a^2 for $a \cdot a$).*

Proof By Order Axiom 3, since $a \neq 0$, either $a \in \mathcal{P}$ or $-a \in \mathcal{P}$.
   Case 1) $a \in \mathcal{P}$. Then since $\mathcal{P}$ is closed under multiplication, $a \cdot a \in \mathcal{P}$.
   Case 2) $-a \in \mathcal{P}$. Then again, since $\mathcal{P}$ is closed under multiplication, $(-a) \cdot (-a) \in \mathcal{P}$. But by Exercise 8.7.4, $(-a) \cdot (-a) = a \cdot a$, so $a \cdot a \in \mathcal{P}$. □

Corollary 8.9.4 *1 is positive.*

Proof $1 \neq 0$ and $1 = 1^2$. □

Corollary 8.9.5 $1 > 0$.

Definition 8.9.6

1. $x \geq y$ means $x > y$ or $x = y$;

2. $x < y$ means $y > x$;

3. $x \leq y$ means $x < y$ or $x = y$.

All the standard properties of these notions are simple consequences of what we have just done. We do not state them since they are rather tedious.

Not every field is an ordered field. For instance, no finite field (and there are many such things) is an ordered field. This is a consequence of Exercise 8.9.14. Further, the complex numbers, though they are a field, are not an ordered field. We do not develop the properties of the complex numbers in this book. But if you are familiar with them, the following argument shows they do not constitute an *ordered* field.

We have shown: 1) in any ordered field, 1 is positive; and 2) in any ordered field, if $a \neq 0$, a^2 is positive, hence $i^2 = -1$ would have to be positive. These two items contradict our third order axiom.

Both the signed reals and the signed rationals are ordered fields, however, and there are other examples as well.

Exercises

Exercise 8.9.1 Show $x > y$ if and only if $a + x > a + y$.

Exercise 8.9.2 Show if $x > y$ and $a > b$ then $x + a > y + b$.

Exercise 8.9.3 Show a is positive if and only if $a > 0$.

Exercise 8.9.4 Show, if $x > y$ and $a > 0$ then $a \cdot x > a \cdot y$.

Exercise 8.9.5 Show $x > y$ if and only if $-y > -x$.

Exercise 8.9.6 (Trichotomy) Show that for any x and y, exactly one of the following is true: $x > y$, $x = y$, $y > x$.

Exercise 8.9.7 Show that in an ordered *field*, if a is positive, it has a multiplicative inverse.

Exercise 8.9.8 Show that in an ordered *field*, if a is positive so is a^{-1}.

Exercise 8.9.9 Show that if $a > 0$, $x > y$ if and only if $a \cdot x > a \cdot y$.

Exercise 8.9.10 Show $a < 0$ if and only if $-a > 0$.

Exercise 8.9.11 Show:

1. if $a > 0$ and $b < 0$ then $a \cdot b < 0$;

2. if $a < 0$ and $b < 0$ then $a \cdot b > 0$.

Exercise 8.9.12 Show that between any two members of an ordered field there is another. Explicitly, show if $a > b$ then $a > \frac{a+b}{1+1} > b$.

Exercise 8.9.13 Show that between any two members of an ordered field there are infinitely many others.

Exercise 8.9.14 Show that, in any ordered field, $1, 1+1, 1+1+1, 1+1+1+1$, etc. are all different. Hint: $1 > 0$. (Similarly for ordered integral domains.)

8.10   The least upper bound property for signed reals

In this section we return to the specific ordered field of signed reals, and we show an analog of the Least Upper Bound Theorem 7.9.7.

Definition 8.10.1 Let $\mathcal{C}$ be a non-empty collection of signed reals.

1. u is an *upper bound* for $\mathcal{C}$ if, for each $x \in \mathcal{C}$, $u \geq x$.

2. u is a *least upper bound* for $\mathcal{C}$ if it is an upper bound, but smaller than any other upper bound.

The notion of least upper bound should not be confused with that of largest. x is the *largest* member of $\mathcal{C}$ if $x \in \mathcal{C}$ and x is bigger than any other member of $\mathcal{C}$. In a similar way we may define the notion of *smallest* member of $\mathcal{C}$. Now, the least upper bound of $\mathcal{C}$ (if it exists) is the smallest of the upper bounds for $\mathcal{C}$. It need not be the largest member of $\mathcal{C}$, since it need not even be a member of $\mathcal{C}$. However, see Exercise 8.10.2.

We now proceed to establish a Least Upper Bound Theorem for the signed reals by, successively: getting it for sets of positive reals, mixed sets of positive and negative reals, and finally arbitrary sets of signed reals. The basic ideas are: first, positive reals act like unsigned reals, and second, negative reals become positive by 'translating them to the right'.

First we show the notion of least upper bound for unsigned reals agrees with the present one for positive reals. That is, temporarily resuming our suppressed notation,

Lemma 8.10.2 *Let C be a set of* unsigned *reals, and let $C^{\mathcal{I}}$ be the corresponding set of positive reals. That is, $\mathcal{I}X \in C^{\mathcal{I}}$ if and only if $X \in C$. Then U is the least upper bound of C if and only if $\mathcal{I}U$ is the least upper bound of $C^{\mathcal{I}}$.*

Proof This follows immediately from two facts:
    1) the notions of least upper bound are defined entirely in terms of $>$ and
    2) by Exercise 8.8.1, $>$ for unsigned reals and $>$ for positive signed reals agree. $\square$

Corollary 8.10.3 *If C is a non-empty set of* positive *reals, bounded from above, then C has a least upper bound.*

Proof Since we have a least upper bound theorem for unsigned reals. $\square$

Theorem 8.10.4 (Least Upper Bound Theorem)
Any non-empty set of signed reals which has an upper bound has a least upper bound.

Proof Let C be a non-empty set of signed reals having an upper bound. Choose any member m of C, and choose any signed real $b < m$.
    Form the collection $C + (-b)$ as defined in Exercise 8.10.5. By part 1 of that exercise, since C has an upper bound, so does $C + (-b)$. Also $C + (-b)$ has some positive members since $m + (-b)$ is in it and $m + (-b) > 0$ since $m > b$.
    Then by the corollary, the *positive* part of $C + (-b)$ has a least upper bound, and by Exercise 8.10.4, this must be a least upper bound for the entire of $C + (-b)$. Finally, Exercise 8.10.5 part 2 gives us that C itself has a least upper bound. This concludes the proof. $\square$

There is a 'dual' version of this which we can get easily now. We leave it to you as Exercise 8.10.7.

Definition 8.10.5 Let C be a non-empty collection of signed reals.

1. b is a *lower bound* for C if, for each $x \in C$, $b \le x$.

2. b is a *greatest lower bound* for C if b is a lower bound, but bigger than any other lower bound.

Exercises

Exercise 8.10.1 Show that if C has a least upper bound, it has only one.

Exercise 8.10.2 Show that if C has a largest member, it is the least upper bound of C.

Exercise 8.10.3 Find a set C of signed reals which has a least upper bound but does not have a largest member.

Exercise 8.10.4 Let C be a set of signed reals, some of which are positive. Let C^+ be the set of positive reals in C and C^o be the rest of C. Show that if u is the least upper bound for C^+, it is also the least upper bound for the entire collection C.

Exercise 8.10.5 Let C be a non-empty collection of signed reals, and let $C + a$ be the collection which results when a is added to each member of C. That is, $C + a = \{x + a \mid x \in C\}$. Show:

1. u is an upper bound for C if and only if $u + a$ is an upper bound for $C + a$;

2. u is the least upper bound for C if and only if $u + a$ is the least upper bound for $C + a$.

Exercise 8.10.6 Show that C can not have more than one greatest lower bound.

Exercise 8.10.7 Show that any non-empty set of signed reals which has a lower bound has a greatest lower bound.

8.11   Complete ordered fields

Suppose $\mathcal{F}$ is an ordered field. It is called *complete* if any non-empty set of members of $\mathcal{F}$ which has an upper bound has a least upper bound. In Section 10 we showed the signed reals were complete. The remarkable fact is that, essentially, there is only one complete ordered field. Any two are intertranslatable. Thus the approach we followed in developing real numbers, using infinite decimals, and any of the other approaches that are commonly used all lead to essentially the same end. This makes possible an elegant characterization of the signed real number system: it is 'the' complete ordered field. These days many books on mathematical analysis start with this as a definition. The only difficulty with this approach is that it does not establish that signed reals exist, it only says how they behave, assuming their existence. In fact, the entire of this book may be looked at as a proof of the following.

Theorem 8.11.1 *Given the assumptions made in the course of this book, a complete ordered field exists.*

This is far enough for this book to go. To continue from here, consult any book on Modern Algebra for the further development of fields, and for exactly what *intertranslatable* means (the technical term is *isomorphic*). Also consult any book on Analysis for the further development of the real number system itself.

"...of making many books there is no end; and much study is a weariness of the flesh."

Ecclesiastes
Ch 12 verse 12
The Bible

"We are like a stray line of a poem, which ever feels that it rhymes with another line and must find it, or miss its own fulfillment. This quest of the unattained is the great impulse in man which brings forth all his best creations...."

Thought Relics
Tagore, 1921

www.ingramcontent.com/pod-product-compliance
Lightning Source LLC
Chambersburg PA
CBHW072302210326
41519CB00057B/2563